Introduction to a Future Way of Thought

Kostas Axelos was a Greek-French philosopher with an emphasis on the studies of Karl Marx and Martin Heidegger. His two doctoral theses were on Heraclitus and Marx. In addition he translated Heidegger, Georg Lukács, and Karl Korsch. He is best known for his works on the concept of the world—particularly the 1969 book *Le jeu du monde*—, as one of the editors of the journal *Arguments*, and as editor of the book series of the same name published by Les Éditions des Minuit.

Introduction to a Future Way of Thought: On Marx and Heidegger

Kostas Axelos

Edited by
Stuart Elden

Translated by
Kenneth Mills

μ meson press

First published in German:
Kostas Axelos: Einführung in ein künftiges Denken: Über Marx und Heidegger © Walter De Gruyter GmbH Berlin/Boston. All rights reserved.

**Bibliographical Information of the
German National Library**
The German National Library lists this publication in the Deutsche Nationalbibliografie (German National Bibliography); detailed bibliographic information is available online at http://dnb.d-nb.de.

Published in 2015 by meson press, Hybrid Publishing Lab,
Centre for Digital Cultures, Leuphana University of Lüneburg
www.meson-press.com

LEUPHANA
CENTRE FOR DIGITAL CULTURES

Design concept: Torsten Köchlin, Silke Krieg
The print edition of this book is printed by Lightning Source, Milton Keynes, United Kingdom.

ISBN (Print):   978-3-95796-005-4
ISBN (PDF):   978-3-95796-006-1
ISBN (EPUB):   978-3-95796-007-8
DOI: 10.14619/009

The digital editions of this publication can be downloaded freely at: www.meson-press.com.

Funded by the EU major project Innovation Incubator Lüneburg

Contents

Introduction: Kostas Axelos— A Left-Heideggerian or a Heideggerian-Marxist

Stuart Elden

Kostas Axelos was born in 1924 in Athens and died in 2010 in Paris. He had left Greece in 1945 at the end of the civil war, under sentence of death from the royalists for his involvement with the communists.[1] Along with Cornelius Castoriadis and Kostas Papaïoannou he was one of those that boarded the ship *Mataroa*.[2] Axelos was fluent in French and German, and chose recently liberated Paris as his new home. Almost immediately on arriving in France he got in touch with the Parti Communiste Français,[3] but found them at once "too Stalinist and too conservative," both politically and in the cultural sphere.[4] Intellectually, France

---

1   On the civil war, see Kostas Axelos, "La guerre civile en Grèce," in *Arguments d'une recherche* (Paris: Éditions de Minuit, 1969), 125–39.

2   Christoph Premat, "A New Generation of Greek Intellectuals in Postwar France," in *After the Deluge: New Perspectives on Postwar French Intellectual and Cultural History of Postwar France*, ed. Julian Bourg (Lanham: Lexington Books, 2004), 103–23.

3   Rémy Rieffel, *La tribu des clercs: Les intellectuels sous la Ve République, 1958–1990* (Paris: Calmann-Lévy / CNRS Éditions, 1993), 290.

4   Kostas Axelos, *"Mondialisation* without the World," interview by Stuart Elden, *Radical Philosophy,* no. 130 (March/April 2005): 25.

at the time was dominated by Marxism and existentialism—a combination Jean-Paul Sartre, among others was trying to realize—yet neither fully satisfied the young Axelos. He found the philosophy of university professors similarly uninspiring.[5] He nonetheless studied at the Sorbonne, and conducted research at the Centre National de la Recherche Scientifique between 1950 and 1957, where he undertook the work that would comprise his primary and secondary theses on Marx and Heraclitus, both later published as books.[6] He returned to the Sorbonne where he taught between 1962 and 1973. His first book of essays was published in Athens in 1952. He became a very important intellectual in postwar France, known not just for his own writings, but also as an editor, translator and interpreter. Axelos had a wide range of intellectual contacts, including Jacques Lacan, Jean Beaufret, and, through them, Martin Heidegger; Pablo Picasso, whose partner Françoise Gilot ended up living with Axelos; Jacques Derrida, Gilles Deleuze, Henri Lefebvre, André Breton and Georges Bataille.[7] He attended seminars by Karl Jaspers and had his thesis examined by, among others, Paul Ricoeur and Raymond Aron.[8] He is cited approvingly by Jacques Derrida in *Of*

5   Ibid.

6   Kostas Axelos, *Marx penseur de la technique: De l'aliénation de l'homme à la conquête du monde*, 2 vols. (1961; Paris: Éditions de Minuit, 1974); translated by Ronald Bruzina as *Alienation, Praxis and Techne in the Thought of Karl Marx* (Austin: University of Texas Press, 1976); *Héraclite et la philosophie: La première saisie de l'être en devenir de la totalité* (Paris: Éditions de Minuit, 1962).

7   Françoise Gilot and Carlton Lake, *Life with Picasso* (New York: McGraw Hill, 1964), 356–57, mentions Axelos briefly. On meeting these figures, see Axelos, "*Mondialisation* without the world," 26.

8   Much of this biographical detail comes from Eric Haviland, *Kostas Axelos: Une vie pensée, une pensée vécue* (Paris: L'Harmattan, 1995), esp. 41 and 65; Henri Lefebvre and Pierre Fougeyrollas, "Notice Bio-Bibliographique," in *Le Jeu de Kostas Axelos* (Montpellier: Fata Morgana, 1973), 97–99; and Ronald Bruzina, "Translator's Introduction," in Axelos, *Alienation, Praxis and Techne*, xxv.

Grammatology and his books were reviewed in *Critique* by Gilles Deleuze and in *Esprit* by Henri Lefebvre.[9]

From 1958 until 1962 he was one of the editors of the very important *Arguments* journal. Edgar Morin and Roland Barthes were also involved, and contributors included Lefebvre, Maurice Blanchot, Gilles Deleuze, and Claude Lefort.[10] Axelos co-translated works by Georg Lukács, introduced translations of Theodor W. Adorno and Karl Korsch, and the journal also translated some of the late writings of Heidegger. The journal was short lived (1956–62) but important in shaping debates for a non-Stalinist left, with links to other journals across Europe: *Ragionamenti* in Italy, *Praxis* in Yugoslavia, *Nowa Cultura* in Poland, and directly inspired the German *Das Argument*.[11] Its links to the British

9   Jacques Derrida, *Of Grammatology*, trans. Gayatri Chakravorty Spivak (Baltimore: Johns Hopkins University Press, 1976), 326, n. 14; Gilles Deleuze, "Faille et feux locaux, Kostas Axelos," review of *Vers la pensée planétaire, Arguments d'une recherche* and *Le jeu du monde*, by Kostas Axelos, *Critique*, no. 275 (April 1970): 344–51; Henri Lefebvre,"Marxisme et technique," review of *Marx penseur de la technique*, by Kostas Axelos, *Esprit*, no. 307 (1962): 1023–28; review of *Vers la pensée planétaire: le devenir-pensée du monde et le devenir-homme [sic] de la pensée,* by Kostas Axelos, *Esprit*, no. 338 (1965): 1114–17. The last is translated in *State, Space, World: Selected Essays,* ed. Neil Brenner and Stuart Elden, trans. Gerald Moore, Neil Brenner and Stuart Elden (Minneapolis: University of Minnesota Press, 2009) 254–58. See also Lefebvre and Fougeyrollas, *Le Jeu de Kostas Axelos*, 97–99; and Henri Lefebvre, "Le monde selon Kostas Axelos," *Lignes*, no. 15 (1992): 129–40; translated as "The World According to Kostas Axelos," in *State, Space, World*, 259–73.

10 All the issues of the journal have been reissued in a two-volume set, with facsimile pages and the addition of short prefaces by Axelos, Edgar Morin, and Jean Duvignaud, as *Arguments 1956–1962: Édition intégrale* (Toulouse: Privat, 1983). All subsequent references are to the issues found in this collection, although the volume number and pages accord to the original issue. For a much more extensive discussion of the journal and the series, see Stuart Elden, "Kostas Axelos and the World of the Arguments Circle," in Bourg, *After the Deluge*, 125–48.

11 Rieffel, *La Tribu des clercs*, 297; Axelos, *Arguments d'une recherche*, 161.

New Left Review were less formal.[12] The journal was closed by its editors, not for lack of money or readers, but because the intellectual project had run its course. The journal ended just as the first books under Axelos's own name had appeared, in 1961 and 1962, on Marx and Heraclitus. The thinkers translated in *Arguments*, and the work published in its pages, would have an important impact on his subsequent career and work: "a great laboratory or fusion of ideas."[13] Axelos stressed that it was not a reflection of a homogenous vision, and the differences between the editorial team and contributors continued throughout. He sketched its key contributions as "an attempt at an open Marxism . . . a revised and corrected Freudo-Marxism and, finally, a post-Marxist and post-Heideggerian thought."[14] And he stressed it was important a journal "that was definitely of the Left was open" to Heidegger, "a great thinker."[15]

By this time, Axelos had also created a book series of the same title with Les Éditions de Minuit, which continued this mode of intellectual plurality. The journal had been an entirely collaborative venture, whereas the book series was the vision of Axelos alone. It stood alongside Georges Bataille's *Critique* series at the forefront of French thought. The *Arguments* book series included texts by Bataille, Beaufret, Blanchot, Deleuze, Lefebvre, René Lourau, Morin, and Didier Franck's important studies of Husserl and Heidegger, as well as almost all of Axelos's own works. Also important, as with the journal, was the program of translations: these comprised Marxists and non-Marxists, often their first works translated into French. Lukács' *Histoire et connaissance de classe* was the first book in the series, appearing in 1960, with

12 See Mariateresa Padova, "Entretien avec Edgar Morin (Paris, 28 September 1978)," in *Studi Francesi* , no. 73 (January–April 1981): 69.

13 Kostas Axelos and Oliver Corpet, "Le fonctionnement," *La revue des revues*, no. 4 (1987): 17; see Axelos, "*Mondialisation* without the world," 26.

14 Axelos, "*Mondialisation* without the world," 27.

15 Ibid.

an important preface by Axelos himself.[16] The English translation only appeared in 1971, and its translator acknowledges the
importance of the French.[17] Works by a range of other thinkers,
including the dissident Marxism of Karl Korsch and Herbert
Marcuse; Louis Hjelmslev and Roman Jakobson's pioneering work
on linguistics; and the phenomenology of Karl Jaspers and Fink
were also included in the series. Axelos also commissioned translations of Eugen Fink's *Spiel als Weltsymbol*—play, or the game, as
symbol of the world—and Wilfrid Desan's *The Planetary Man*, to
which Axelos contributed an afterword.[18] Fink was Husserl's assis-
tant for many years and co-organized with Heidegger a seminar
on Heraclitus.[19] His work, principally on the world, was profoundly
important to Axelos, especially regarding their shared project of
elaborating Heraclitus's famous fragment that declared time, the
world, or the universe was "like a child playing a game."[20] Axelos
believed that unorthodox or open Marxism could be reconciled
with Heidegger and the post-Heidegger thought of figures like
Fink. His attempt at a left-Heideggerianism or a Heideggerian
Marxism is found throughout his works but in two places above
all: his book on Marx and technology, and in this book, *On Marx
and Heidegger.*

16   Kostas Axelos, "Preface de la presente édition," in Georg Lukács, *Histoire et
      connaissance de classe: essais de dialectique marxiste*, trans. Kostas Axelos
      and Jacqueline Bois (Paris: Éditions de Minuit, 1960), 1–8.

17   Georg Lukács, *History and Class Consciousness*, trans. Rodney Livingstone
      (London: Merlin, 1971).

18   Eugen Fink, *Le Jeu comme symbole du monde*, trans. Hans Hildenbrand
      and Alex Lindenberg (Paris: Éditions de Minuit, 1966); Wilfrid Desan, *The
      Planetary Man: A Noetic Prelude to a United World* (Washington: Georgetown
      University Press, 1961) and *L'Homme planétaire: prélude théorique à un
      monde uni*, trans. Hans Hildenbrand and Alex Lindenberg (Paris: Éditions de
      Minuit, 1968); Axelos's "Postface: Qui est donc l'homme planétaire," 151–57,
      is reprinted in *Arguments d'une recherchere*, 181–86. Axelos discusses Fink in
      this text.

19   Martin Heidegger and Eugen Fink, *Heraclitus Seminar 1966/67*, trans. Charles
      E. Seibert (Tuscaloosa: University of Alabama Press), 1979.

20   See Stuart Elden, "Eugen Fink and the Question of the World," *Parrhesia*, no.
      5 (2008): 48–59.

Axelos is little known in English-speaking contexts, partly due to a lack of translations. Before this current volume, the only book translated into English had been the one on Marx and technology and a small number of essays. The situation is very different in other languages, with his works available in multiple translations, especially in Spanish and Italian. He wrote most of his books in French, with four in Greek and one—this volume—in German. There is also a lack of secondary literature in English. Mark Poster's *Existential Marxism in Postwar France* (1975) and Ronald Bruzina's introduction to his translation of *Alienation, Praxis, and Techne in the Thought of Karl Marx* provide the most substantial treatment.[21] My own *Understanding Henri Lefebvre* traces some of the interrelations between Lefebvre and Axelos.[22]

Reading Marx and Heidegger

There are a number of crucial themes in Axelos's appropriation of Marx. As the references in this volume attest, he read Marx from early to late—from the 1844 "Economic and Philosophical Manuscripts" to *Capital*, the *Grundrisse* and the "Critique of the Gotha Programme." He found the early writings crucial for their analysis of alienation and the critique of political economy to be a deepening of these claims. Axelos shows how alienation, a major concern of Marx, especially in his early writings, is related to Heidegger's notion of the "forgetting of being."[23] Alienation in Marx, according to Axelos, can be found not only in ideology

21   Mark Poster, *Existential Marxism in Postwar France* (Princeton: Princeton University Press, 1975); Bruzina, "Translator's Introduction," in Axelos, *Alienation, Praxis and Techne*, ix–xxxiii.

22   Stuart Elden, *Understanding Henri Lefebvre: Theory and the Possible* (London: Continuum, 2004); "Lefebvre and Axelos: *Mondialisation* before Globalisation," in *Space, Difference, Everyday Life: Reading Henri Lefebvre*, ed. Kaniska Goonewardena, Stefan Kipfer, Richard Milgrom, and Christian Schmid (New York: Routledge, 2008), 80–93.

23   Kostas Axelos and Dominique Janicaud, "Entretiens du 29 janvier 1998 et du mars 2000," in Dominique Janicaud, *Heidegger en France* (Paris: Albin Michel, 2001), 2: 11.

and economics but in technology as well. Marx's own writings examine this most thoroughly in the labor process and the circulation of capital. Axelos reads Marx widely to investigate this issue, including lesser-examined works, such as *The German Ideology* and *The Poverty of Philosophy*. In Axelos's reading, pushing Marx further, it is technology that challenges the way in which we deal with nature, the world, and the entirety of our social relations. This examination of a Heideggerian question in Marx both provides a new way to analyze Marx and provides a more progressive politics to Heideggerian inquiries. It would then characterize much of his work from the early study of Marx to the current volume and through to many of his later works.

Axelos describes modern technology as an *échafaudage*, a scaffold or a framework,[24] and thinks this has become the "*worldwide technical* échafaudage."[25] Such a description becomes clearer when we recognize that "échafaudage" is the term Axelos suggests best translates the Heideggerian notion of *das Ge-stell*, usually translated as "en-framing," or, in French, as *arraissonment* or *dispositif*.[26] Axelos, building on Heidegger, suggests that the way that we conceive of the world is founded upon a particular ontological determination, conceiving of the world as calculable, measurable, and therefore controllable and exploitable. This determination, coming out of European modernity, leads to the present planetary era. For Axelos, "[t]his era is global and world-wide, errant, planing and flattening, planning, calculating and combinative [*Cette ère est globale et mondiale, errante, aplanissante*

---

24 See, for example, Kostas Axelos, *Problèmes de l'enjeu* (Paris: Éditions de Minuit, 1979), 16 and 66.

25 Kostas Axelos, *Contribution à la logique* (Paris: Éditions de Minuit, 1977), 80, see 121.

26 Kostas Axelos, note to Heidegger, "Principe d'identité," trans. Gilbert Kahn, *Arguments*, no. 7 (1958): 5, n. 1. Kahn, the translator of the piece, uses *la com-mande*. For a brief discussion of the term in French, see Stuart Elden, *Mapping the Present: Heidegger, Foucault and the Project of a Spatial History* (London: Continuum, 2001), 110–11.

et aplatissante, planificatrice, calculatrice et combinatoire]."[27]
Much of the terminology in this formulation will appear in the
present text. In particular, Axelos is interested in the idea of
errancy, wandering or error. He regularly picks up on the Greek
description of a planet as a wandering star—*aster planétes*—
and uses this to describe both the world and humanity. For
Axelos, following Heidegger, the framework that makes modern
technology achievable precedes it as its condition of possibility.
Marx's critique of political economy is based upon trying to
comprehend the reduction of phenomena to value—use or
exchange—a numerical measure of productivity and power. By
technology [*technique*] Axelos means something much broader
than tools or techniques. Rather he is concerned with inter-
rogating their underlying logic.

> Technology founds, undoubtedly, the possibility and
> effectiveness of machines, industry, the exploitation of
> atomic energy and of all other energy, but it goes far beyond
> apparatuses and machinery. And it is planetary technology
> which orders the new worldwide politics, the planetary
> politics.[28]

But Axelos, as well as finding a Heideggerian problematic developed in Marx, reads Marx in much the same way as Heidegger
reads Nietzsche, as the final figure of Western metaphysics, in
whom the most radical challenge and the exhaustion of possibilities comes together.[29] In his own writings, Heidegger only
briefly discusses Marx, and the role that Marx plays in the final
stages of metaphysics—Axelos discusses the key passages in

27 Axelos, *Arguments d'une recherche*, 174. In *Horizons du monde* (Paris: Éditions
    de Minuit, 1974), 112–13, this notion is linked to the idea of the end of history.
    On technology generally, see the essays in Kostas Axelos, *Métamorphoses:
    Clôture–Ouverture* (Paris: Éditions de Minuit, 1991),

28 Kostas Axelos, *Vers la pensée planétaire: Le devenir-pensée du monde et le
    devenir-monde de la pensée* (Paris: Éditions de Minuit, 1964), 297.

29 Axelos, *Arguments d'une recherche*, 168; *Marx penseur de la technique*, 2:
    120–21; *Alienation, Praxis, and Techne*, 246.

this work. It is interesting to note that the principal places where Heidegger does deal with this theme are in relation to French promptings—in the "Letter on Humanism" to Jean Beaufret, and in the "What is Philosophy?" lecture at Cerisy-la-Salle: an event for which Axelos acted as interpreter. When Axelos suggested that Heidegger has not sufficiently dealt with Marx's insights into technology, or indeed Marx at all, Heidegger replied that he should undertake this work himself.[30] In this present work, and in the book on Marx and technology, we find Axelos doing precisely that. In addition, like Heidegger, Axelos turns to pre-Socratics thinkers, especially Heraclitus. As well as the book on Heraclitus he translated and edited a collection of the available fragments with commentary.[31] Lefebvre even went so far as to describe Axelos as the "new Heraclitus,"[32] because of his dialectical, historical way of thinking. Lefebvre additionally praised Axelos as the most significant thinker to have grasped Heraclitus's teaching of thought *of* the world and thought *in* the world.[33]

On World

The concept of "world," *le monde*, and processes of its thinking and transformation had been discussed extensively in the *Arguments* journal. A key early text was "Thèses sur la mondialisation," by Pierre Fougeyrollas in 1959. Here it was suggested that "to the *mondialisation* of problems we must respond with the *mondialisation* of thought and action."[34] A subsequent issue of

30   Axelos and Janicaud, "Entretiens," 15; Haviland, *Kostas Axelos*, 56.

31   Kostas Axelos, *Héraclite et la philosophie*; *Les Fragments d'Héraclite d'Ephèse*, edited and translated by Kostas Axelos (Paris: Éditions Estienne, 1958). For a similar agenda by another member of the *Arguments* group, see François Châtelet, *Logos et praxis: Recherches sur la signification théorique du Marxisme* (Paris: Société d'Édition d'Énseignement Supérieur, 1962).

32   Lefebvre, "Au-delà du savoir," in Lefebvre and Fougeyrollas, *Le jeu de Kostas Axelos*, 32.

33   Henri Lefebvre, *Qu'est-ce que penser?* (Paris: Publisad, 1986), 13; see also "Au-delà du savoir," 24–26.

34   Pierre Fougeyrollas, "Thèses sur la mondialisation," *Arguments*, no. 15 (1959): 38–39.

the journal devoted a theme section to the topic of "the planetary era"; and in their 1960 manifesto, the editors suggested that one of its explicit purposes was to make sense of the "second half of the twentieth century: a planetary age of technology; an iron age of a new industrial civilization; a new age of the human."[35] In addition one of the two divisions of the *Arguments* book was entitled "The Becoming-Thought of World and the Becoming-Worldly of Thought."[36] This phrase picks up on a key claim in Marx's doctoral thesis: "the world's becoming philosophical is at the same time philosophy's becoming worldly, that its realization is at the same time its loss."[37] Axelos regularly cited this as an aphorism and it could be said to serve as a guiding theme for his entire work.[38] Marx's point is that in its becoming worldly, that is in its actualization, philosophy is transcended and overcome. What is interesting here, in relation to the book series, is that "philosophy" is replaced by "thought"—another Marxist theme transfigured through Heidegger.

Mondialisation is the process of becoming-worldly, something which relates to the Anglophone term "globalization" but cannot be simply equated with it. The argument of Axelos and his journal collaborators is that the 1960s saw a new era of planetary technology and *mondialisation*.[39] They put the emphasis on the process of becoming worldly, the seizing and comprehending of the world as a whole, as an event in thought, rather than on the spread of phenomena of economics and politics across the surface of the globe. In other words, the second process, globalization, is in a sense only possible because of this prior

35   "Manifeste no. 2 (1960)," in Axelos, Morin, and Duvignaud, *Arguments 1956–1962*, xxx.

36   Axelos, *Arguments d'une recherche*, 164–65. For a discussion, see *Vers la Pensée planétaire*, 13, 30.

37   Karl Marx, *Writings of the Young Marx on Philosophy and Society*, ed. Loyd D. Easton and Kurt H. Guddat (New York: Doubleday, 1967), 62.

38   Axelos, *Marx penseur de la technique*, 1: 4, 2: 50, 162; *Alienation, Praxis and Techne*, v, 202, 271; *Problèmes de l'enjeu*, 177; and the present volume.

39   See Axelos, *Arguments d'une recherche*, 162.

comprehending of the world, *mondialisation*. Although the distinction between the two terms has been blurred in more recent French writings, it is important in understanding the concepts in their usage at the time. This issue is explored in much detail in Axelos's writings, and was developed, for example, in Lefebvre's work on the state and production on the world scale.[40] These are some of the earliest usages of the term in French literature, and predate the discussion of the notion of globalization in English-language scholarship. As Axelos suggested much later, when globalization was much more widely discussed, the term globalization—affecting the globe—misses the "world" and so-called world history.[41]

> *Globalization* names a process which universalizes technology, economy, politics, and even civilization and culture. But it remains somewhat empty. The world as an *opening* is missing. The world is not the physical and historical totality, it is not the more or less empirical ensemble of theoretical and practical ensembles. It deploys itself. The thing that is called *globalization* is a kind of *mondialisation* without the world.[42]

The issue of the world, particularly in relation to the notion of play or the game—*le jeu*—was a recurrent concern of Axelos's own writings. Axelos would pick up on Heidegger's suggestion that "the essence of being is the game itself [*das Spiel selber*]."[43] In numerous works, notably *Vers la Pensée planétaire*, *Horizons du monde*, *Le Jeu du monde*, and the third part of the present volume,

---

40 See, in particular, Henri Lefebvre, *De l'État*, 4 vols. (Paris: UGE), 1976–78; some of the chapters in *State, Space, World,* and for a commentary, Elden, *Understanding Henri Lefebvre*, chap. 6.

41 Kostas Axelos, *Ce questionnement: Approche-Éloignement* (Paris: Éditions de Minuit, 2001), 40.

42 Axelos, "*Mondialisation* without the world," 27.

43 Martin Heidegger, "Identität und Differenz," in *Gesamtausgabe*, vol. 11, *Identität und Differenz*, ed. Friedrich-Wilhelm von Herrmann (Frankfurt am Main: Vittorio Klostermann, 2006), 72. See also Axelos, *Vers la Pensée planétaire*, 22.

Axelos developed his thoughts on this question. The major work is *Le Jeu du monde*, which took him fifteen years to write, but there the ideas are presented in difficult, aphoristic style across more than four hundred pages without references or discussion of related thinkers. His 1984 book, *Systématique ouverte*, one chapter of which is available in English translation, provides a much more approachable distillation of his ideas.[44]

For Axelos therefore:

> The world cannot be reduced either to an ensemble of intraworldly phenomena, nor to "creation," or to the Cosmic Universe, to which is adjoined a social and historical world, nor to the totality of that which human representation understands, nor to the total scope of technical activity.[45]

The work on the world and on technology are closely inter-related. Axelos suggests that talk of an "atomic" or "nuclear era" at the time he was writing was not sufficiently thought through. What was actually being named? For him, both designations are symptoms of the much wider context of planetary technology, a new destiny of the world.[46] Axelos describes this as the "becoming-worldly of technology, and the becoming-technological of the world," in another twist to the Marxist phrase.[47] The crisis facing humanity has therefore developed: if Nietzsche discussed *European* nihilism,[48] which Heidegger

44 Kostas Axelos, *Systématique ouverte* (Paris: Éditions de Minuit, 1984), part translated by Gerald Moore as "The World: Being Becoming Totality," *Environment and Planning D: Society and Space* 24, no. 5 (2006): 643–51.

45 Axelos, *Problèmes de l'enjeu*, 30.

46 On this in detail see "La question de la technique planétaire," in Axelos, *Ce Questionnement*, 15–35.

47 Axelos, *Métamorphoses*, 132.

48 See especially, Martin Heidegger, GA, vol. 48, *Nietzsche: Der europäische Nihilismus*. A shorter version is translated as *Nietzsche*, vol. 4, *Nihilism,* ed. David Farrell Krell, trans. Frank A. Capuzzi (San Francisco: Harper Collins, 1991).

analyzed in his final lecture course on Nietzsche, Axelos now saw **21** nihilism on the worldwide level.[49]

In this mode of thinking, the term "world" does not simply signify the whole, or the totality of all that exists; it is concerned with relations, interplay, and the game—*le jeu*.[50] For Axelos, following Heidegger, while we cannot equate the world and the human, neither should we think of them entirely separately: "Neither of them is the other, but they cannot act [*jouer*] without the other."[51] Heidegger calls this 'being-in-the-world', which should not be understood in a primarily spatial sense, but rather as an integration of the human and the environment. Axelos underlines this point: "there is not the human *and* world. The human is not *in* the world."[52] What this means is that we are not so much in the world but *of* the world, just as the world is not in space-time, but is spatiotemporal.[53] Our relation with the world is the crucial issue; it is both something within and outside our control:

> The *world* deploys itself as a *game*. That means that it refuses any sense, any rule that is exterior to itself. The play *of* the world itself is different from all the particular games that are played *in* the world. Almost two-and-a-half thousand years after Heraclitus, Nietzsche, Heidegger, Fink and I have insisted on this approach to the world as game.[54]

Pierre Fougeyrollas, "Au-delà du nihilisme," in Lefebvre and Fougeyrollas, *Le jeu de Kostas Axelos*, 77.

50   Kostas Axelos, *Lettres à un jeune penseur* (Paris: Éditions de Minuit, 1996), 10, n. 1, 13.

51   Ibid., 13.

52   Axelos, *Ce Questionnement*, 56.

53   Axelos, *Lettres à un jeune penseur*, 19. See also Axelos, *Systématique ouverte*, 40–54. This chapter is translated by Gerald Moore as "The World: Being Becoming Totality."

54   Axelos, "*Mondialisation* without the world," 28.

The human is the great partner of the play of the world, yet the human is not only the player, but is equally the "out-played" [*déjoué*], the plaything [*jouet*].[55]

—

On Marx and Heidegger is an unusual book in Axelos's overall trajectory. It comprises texts originally composed in German and some of his own translations of texts originally published in French. The major, first part of the book is on the Marx and Heidegger relation, a neglected topic that Axelos is one of the very first to examine in detail.[56] The second part of the book comprises a sequence of "Theses on Marx," which build upon and parody Marx's own "Theses on Feuerbach" (the title is Engels's) and a text discussing the concept of world in Heidegger. This interest in the world, or the 'planetary' leads to the third part of the book, which discusses technology, praxis and science.

The combination of Marx and Heidegger was, and continues to be, provocative. Axelos was a Marxist in thought and a communist in political action, and the work in this book was begun barely a decade after Heidegger had been banned from teaching because of his political actions under the Nazi party. Axelos knew Heidegger, acting as his interpreter for the Cerisy-la-Salle lecture in 1955—and co-translator, with Beaufret, of the published text, and staying with Heidegger in the Black Forest hut.[57] Over the next few years, Axelos also acted as interpreter for Heidegger's meetings with René Char and Georges Braque, and spent several

55 Kostas Axelos, *Entretiens, réels, imaginaires et avec soi-même* (Montpellier: Fata Morgana, 1973), 53.

56 See now Laurence Paul Hemming, *Heidegger and Marx: A Productive Dialogue over the Language of Humanism* (Evanston: Northwestern University Press, 2013).

57 Martin Heidegger, "Qu'est-ce que la philosophie?," trans. Kostas Axelos and Jean Beaufret, in *Questions I et II* (Paris: Gallimard, 1968), 317–46. For an account of his Christmas 1955 visit, see *Vers la Pensée planétaire*, 224–25.

days at Lacan's country house in the company of Beaufret,
Heidegger, and Lacan.[58]

Axelos was no uncritical Heideggerian. While he made extensive use of his ideas, and—especially in *On Marx and Heidegger*—aspired to write like him, he did not shy away from the political aspects of Heidegger's career. He repeatedly questioned Heidegger about his allegiance to the Nazi party, but claims he never got much beyond straightforward explanations. As he said in 2004:

> The discussion of the political question with Heidegger never advanced very far. One must say, the political realm in general eluded him. He was a great thinker and a narrow-minded petty bourgeois at the same time; he did not really understand what had happened and was happening on this level. In the discussions, he tried to exonerate himself, saying that he had committed a great error, that in the beginning National Socialism was not what it later became, that he had distanced himself from Nazism, and so on. All this was wholly insufficient. But despite the National Socialist enticement of Heidegger, his thought can absolutely not be reduced or limited to Nazism. It is an opening, but it remains covered by a shadow. This shadow cannot and must not be forgotten, but all reductive attempts to explain it fail entirely.[59]

Heidegger would claim to Axelos that "in my work, there is no trace of Nazism." Yet even on the basis of materials available at that time Axelos knew that this was clearly false.[60] Newly published materials have confirmed this impression, though they undoubtedly make it more explicit. Yet as early as 1959, Axelos, Beaufret, François Châtelet, and Lefebvre had debated numerous aspects of Heidegger's work, including his relation to Marx and

58   Axelos and Janicaud, "Entretiens," 12; Haviland, *Kostas Axelos*, 50.

59   Axelos, *"Mondialisation* without the world," 26.

60   Axelos and Janicaud, "Entretiens," 14; see Haviland, *Kostas Axelos*, 52–53.

his Nazi past.[61] This should give the lie to any suggestion that French Heidegger scholarship neglected this question until Victor Farías's book in 1987: the question had also been discussed in *Les Temps modernes* as early as the 1940s.[62] Axelos thought you could be a Heideggerian without being on the political Right, but that this could not be at the expense of a detailed and careful interrogation of the relationship between his politics and his thought.[63]

As with most of Axelos's books, *On Marx and Heidegger* is a whole composed of fragments. Only a few of his books have a stronger narrative arc than this. The key exceptions are his early studies of Marx and Heraclitus and his masterwork *Le jeu du monde* [The play (or game) of the world]. Several other books are collections of essays or lectures—*Horizons du monde, Vers une pensée planétaire: Le devenir-pensée du monde et de devenir-monde de la pensée, Arguments d'une recherche.*

Axelos viewed most of his output in terms of trilogies. He discusses the relation between them in a number of places, particularly in *Problèmes de l'enjeu.*[64] The books on Marx and Heraclitus, his primary and secondary theses, were partnered by *Vers la pensée planétaire* in the first trilogy. His works on logic

61   Kostas Axelos, Jean Beaufret, François Châtelet, and Henri Lefebvre, "Karl Marx et Heidegger," in Axelos, *Arguments d'une recherche*, 93–105; originally published in *France Observateur*, no. 473, May 28, 1959. See also *Vers la Pensée planétaire*, 223–25.

62   Victor Farías, *Heidegger and Nazism*, trans. Paul Burrell and Gabriel R. Ricci (Philadelphia: Temple University Press), 1989. See Axelos and Janicaud, "Entretiens," 15. Karl Löwith, "Les implications politiques de la philosophie de l'existence chez Heidegger," *Les Temps modernes*, no. 14 (November 1946): 343–60.

63   Axelos, *Vers la Pensée planétaire*, 224. See also *Métamorphoses*, 16–17, n. 1. This is a position shared with some more recent interventions. See Dominique Janicaud, *L'Ombre de cette pensée* (Grenoble: Jérome Millon, 1990); and my own *Speaking Against Number: Heidegger, Language and the Politics of Calculation* (Edinburgh: Edinburgh University Press, 2006).

64   Axelos, *Problèmes de l'enjeu*, 188–90. For an overall schema of his work, see Haviland, "Le déploiement de l'œuvre," in *Kostas Axelos*, 77–134.

and ethics were the first and third volumes of the second trilogy,
joining *Le jeu du monde*. The final trilogy comprised *Arguments d'une recherche*, *Horizons du monde*, and *Problèmes de l'enjeu*. Each of these trilogies were given an overall title: the unfolding, unfurling, or deployment [*deploiement*] of *errance*, of the game, and of an inquiry. *Le Jeu du monde* is therefore the central book in the central trilogy.

The Trilogy of Trilogies—Axelos's Principal Writings

Le déploiement de l'errance	Le déploiement du jeu	Le déploiement d'une enquête
Héraclite et la philosophie (1962)	*Contribution à la logique* (1977)	*Arguments d'un recherche* (1969)
Marx penseur de la technique (1961)	*Le jeu du monde* (1969)	*Horizons du monde* (1974)
Vers la pensée planétaire (1964)	*Pour une éthique problématique* (1972)	*Problèmes de l'enjeu* (1979)

On Marx and Heidegger sits within the time-frame of these studies, being composed between 1957 and its 1966 publication, and discusses related themes. One short piece also appears in *Vers la pensée planétaire*, but otherwise the overlap is only thematic. Axelos describes the first trilogy as a certain grasp of the play [*saisie du jeu*] of the history of thought and the thought of history; and the second trilogy as presenting, without representing, "a systematic of thought: a logic and a methodology; a questioning and encyclopedic ontology, fundamental, and animating regional ontologies; an anthropology and an ethic."[65] These are major themes and claims, and the three trilogies and his other writings are extremely self-referential and have the impression of an almost Kantian architectonic. Yet they are presented in such a way to resist the idea that he was building a system, but rather an "open systematic."[66] Within this the

---

65   Axelos, *Contribution à la logique*, 7.
66   See Axelos, *Systématique ouverte*, especially 36.

overarching theme had been the question of the world, the play or the game of the world, and the relation of the human to that world of which they are both part and creator.

Axelos is therefore extremely important in terms of his network of contacts and because of the way in which he brought into print a range of texts showing disparate interests. His importance as a facilitator of translation alone is worthy of note. In his own writings the principal interlocutors are Heraclitus, Marx, and Heidegger. Other figures—Hegel and Nietzsche for example—are mentioned, and there are studies of Pascal, Freud, and Rimbaud, but these three are the central ones.[67] *On Marx and Heidegger* gives an excellent introduction to the way he thinks *about* these two thinkers: with them, through them, and *beyond* them.[68] It therefore opens up a route to understanding his work more generally, especially his still timely thought of the world, the planetary, of which the third section of this volume gives some initial indications.

Bibliography

Axelos, Kostas. *Arguments d'une recherche*. Paris: Éditions de Minuit, 1969,

———. *Ce questionnement: approche-éloignement*. Paris: Éditions de Minuit, 2001.

———. *Contribution à la logique*. Paris: Éditions de Minuit, 1977.

———. *Entretiens, réels, imaginaires et avec soi-même*. Montpellier: Fata Morgana, 1973.

———. "La guerre civile en Grèce." In *Arguments d'une recherche*, 125–39. Paris: Éditions de Minuit, 1969.

———. *Héraclite et la philosophie: la première saisie de l'être en devenir de la totalité*. Paris: Éditions de Minuit, 1962.

———. *Horizons du monde*. Paris: Éditions de Minuit, 1974.

———. *Lettres à un jeune penseur*. Paris: Éditions de Minuit, 1996.

———. *Marx penseur de la technique: De l'aliénation de l'homme à la conquête du monde*. 2 vols. 1961. Paris: Éditions de Minuit, 1974.

---

67 In Axelos, *Horizons du monde*, 93, he talks of the "constellation Hegel-Marx-Nietzsche-Freud-Heidegger."

68 See the opening lines of his foreword to this volume for a discussion of the multiple meanings of *über*.

Alienation, Praxis and Techne in the Thought of Karl Marx. Translated by Ronald Bruzina. Austin: University of Texas Press, 1976.

———. *Métamorphoses: Clôture-Ouverture.* Paris: Éditions de Minuit, 1991.

———. *"Mondialisation* without the World." Interview by Stuart Elden. *Radical Philosophy,* no. 130 (March/April 2005): 25–28.

———. "Postface: Qui est donc l'homme planétaire." In Wilfrid Desan. *L'Homme planétaire: prélude théorique à un monde uni.* Translated by Hans Hildenbrand and Alex Lindenberg, 151–57. Paris: Éditions de Minuit, 1968.

———. "Preface de la presente édition." In Georg Lukács. *Histoire et connaissance de classe: essais de dialectique marxiste.* Translated by Kostas Axelos and Jacqueline Bois, 1–8. Paris: Éditions de Minuit, 1960.

———. "Le principe d'identité." Translated by Gilbert Kahn. *Arguments,* no. 7 (1958): 2–8.

———. *Problèmes de l'enjeu.* Paris: Éditions de Minuit, 1979.

———. *Systématique ouverte.* Paris: Éditions de Minuit, 1984.

"The World: Being Becoming Totality." *Environment and Planning D: Society and Space* 24, no. 5 (2006): 643–51.

———. *Vers la Pensée planétaire: le devenir-pensée du monde et le devenir-monde de la pensée.* Paris: Éditions de Minuit, 1964.

Axelos, Kostas, Jean Beaufret, François Châtelet, and Henri Lefebvre. "Karl Marx et Heidegger." in Axelos. *Arguments d'une recherche,* 93–105; originally published in *France Observateur,* no. 473, May 28, 1959.

Axelos, Kostas, and Oliver Corpet. "Le fonctionnement." *La revue des revues,* no. 4 (1987): 15–17.

Axelos, Kostas, and Dominique Janicaud. "Entretiens du 29 janvier 1998 et du mars 2000." In Dominique Janicaud. *Heidegger en France.* 2 vols, 2, 11–33. Paris: Albin Michel, 2001.

Axelos, Kostas, Edgar Morin, and Jean Duvignaud. *Arguments 1956–1962:* Édition *intégrale.* Toulouse: Privat, 1983.

Bourg, Julian, ed. *After the Deluge: New Perspectives on Postwar French Intellectual and Cultural History of Postwar France.* Lanham: Lexington Books, 2004.

Bruzina, Ronald. "Translator's Introduction." In Axelos, *Alienation, Praxis and Techne,* ix–xxxiii.

Châtelet, François. *Logos et praxis: recherches sur la signification théorique du Marxisme.* Paris: Société d'Édition d'Énseignement Supérieur, 1962.

Deleuze, Gilles. "Faille et feux locaux, Kostas Axelos." Review of *Vers la pensée planétaire, Arguments d'une recherche* and *Le jeu du monde,* by Kostas Axelos. *Critique,* no. 275 (April 1970): 344–51.

Derrida, Jacques. *Of Grammatology.* Translated by Gayatri Chakravorty Spivak. Baltimore: Johns Hopkins University Press, 1976.

Desan, Wilfrid. *The Planetary Man: A Noetic Prelude to a United World.* Washington: Georgetown University Press, 1961.

L'Homme planétaire: prélude théorique à un monde uni. Translated by Hans Hildenbrand and Alex Lindenberg. Paris: Éditions de Minuit, 1968.

28   Elden, Stuart. "Eugen Fink and the Question of the World." *Parrhesia*, no. 5 (2008): 48–59.

———. "Kostas Axelos and the World of the Arguments Circle." In Bourg, *After the Deluge*, 125–48.

———. "Lefebvre and Axelos: *Mondialisation* before Globalisation." In *Space, Difference, Everyday Life: Reading Henri Lefebvre*. Edited by Kaniska Goonewardena, Stefan Kipfer, Richard Milgrom, and Christian Schmid, 80–93. New York: Routledge, 2008.

———. *Mapping the Present: Heidegger, Foucault and the Project of a Spatial History*. London: Continuum, 2001.

———. *Speaking Against Number: Heidegger, Language and the Politics of Calculation*. Edinburgh: Edinburgh University Press, 2006.

———. *Understanding Henri Lefebvre: Theory and the Possible*. London: Continuum, 2004.

Farías, Victor. *Heidegger and Nazism*. Translated by Paul Burrell and Gabriel R. Ricci, Philadelphia: Temple University Press, 1989.

Fink, Eugen. *Le Jeu comme symbole du monde*. Translated by Hans Hildenbrand and Alex Lindenberg. Paris: Éditions de Minuit, 1966.

Fougeyrollas, Pierre. "Thèses sur la mondialisation." *Arguments*, no. 15 (1959): 38–39.

Gilot, Françoise, and Carlton Lake. *Life with Picasso*. New York: McGraw Hill, 1964.

Haviland, Eric. *Kostas Axelos: une vie pensée, une pensée vécue*. Paris: L'Harmattan, 1995.

Heidegger, Martin. "Identität und Differenz." In: *Identität und Differenz: Gesamtausgabe Band 11*. Edited by Friedrich-Wilhelm von Herrmann, 27–110. Frankfurt am Main: Vittorio Klostermann, 2006.

———. *Nietzsche: Der europäische Nihilismus*. GA. Vol. 48.
        Nietzsche. Vol. 4, *Nihilism*. Edited by David Farrell Krell. Translated by Frank A. Capuzzi. San Francisco: Harper Collins, 1991.

———. "Qu'est-ce que la philosophie?" In *Questions I et II*. Translated by Kostas Axelos and Jean Beaufret, 317–47. Paris: Gallimard, 1968.

Heidegger, Martin, and Eugen Fink. *Heraclitus Seminar 1966/67*. Translated by Charles E. Seibert. Tuscaloosa: University of Alabama Press, 1979.

Hemming, Laurence Paul. *Heidegger and Marx: A Productive Dialogue over the Language of Humanism*. Evanston: Northwestern University Press, 2013.

Heraclitus. *Les Fragments d'Héraclite d'Ephèse*. Edited and translated by Kostas Axelos. Paris: Éditions Estienne, 1958.

Janicaud, Dominique. *L'Ombre de cette pensée*. Grenoble: Jérome Millon, 1990.

Lefebvre, Henri. "Au-delà du savoir." In Henri Lefebvre and Pierre Fougeyrollas. *Le jeu de Kostas Axelos*. Montpellier: Fata Morgana, 1973.

———. *De l'état*. 4 vols. Paris: UGE, 1976–78.

———. "Marxisme et technique." Review of *Marx penseur de la technique: De l'aliénation de l'homme à la conquête du monde*, by Kostas Axelos. Esprit, no. 307 (1962): 1023–1028.

———. *Qu'est-ce que penser?* Paris: Publisad, 1986.. "Le monde selon Kostas Axelos." *Lignes*, no. 15 (1992): 129–40.

"The World According to Kostas Axelos." In *State, Space, World*, 259–73.

———. *Qu'est-ce que penser?* Paris: Publisad, 1986.

———. Review of *Vers la pensée planétaire: le devenir-pensée du monde et le devenir-homme* [sic] *de la pensée*, by Kostas Axelos. *Esprit*, no. 338 (1965): 1114–17.

———. *State, Space, World: Selected Essays*. Edited by Neil Brenner and Stuart Elden. Translated by Gerald Moore, Neil Brenner and Stuart Elden. Minneapolis: University of Minnesota Press, 2009.

Lefebvre, Henri, and Pierre Fougeyrollas. *Le jeu de Kostas Axelos*. Montpellier: Fata Morgana, 1973.

Löwith, Karl. "Les implications politiques de la philosophie de l'existence chez Heidegger." *Les Temps modernes*, no. 14 (November 1946): 343–60.

Lukács, Georg. *History and Class Consciousness*. Translated by Rodney Livingstone. London: Merlin, 1967.

"Manifeste no. 2 (1960)." In Axelos, Morin, and Duvignaud, *Arguments 1956–1962*, xxx.

Marx, Karl. *Writings of the Young Marx on Philosophy and Society*. Edited by Loyd D. Easton and Kurt H. Guddat. New York: Doubleday, 1967.

Padova, Mariateresa. "Entretien avec Edgar Morin (Paris, 28 September 1978)." In *Studi Francesi*, no. 73 (January–April 1981): 46–72.

Poster, Mark. *Existential Marxism in Postwar France*. Princeton: Princeton University Press, 1975.

Premat, Christoph. "A New Generation of Greek Intellectuals in Postwar France." In Bourg, *After the Deluge*, 103–23.

Rieffel, Rémy. *La tribu des clercs: les intellectuels sous la Ve République 1958–1990*. Paris: Calmann-Lévy / CNRS Éditions, 1993.

Notes on the Translation

The following work occupies a relatively early position in the—French and German—*oeuvre* of Kostas Axelos, although many works still await translation into the English language.

Even here, the importance of the key concepts "game" and "errance"—indispensable for an understanding of his philosophical ductus and work as a whole—is evident.

Although continuity has been sought for this translation in respect to A*lienation, Praxis, and Techne in the Thought of Karl Marx* by Axelos, translated by Ronald Bruzina (University of Texas Press, Austin & London, 1976), this only proved to be necessary to a very limited degree and the following translation, including quotations from Nietzsche, Heidegger, and others, are new.

As regards translations of Heideggerian terminology, the solution used here also seeks a certain continuity with the dedicated and seminal work of Theodore Kisiel (*The Genesis of Heidegger's Being and Time*, University of California Press, Berkeley/Los Angeles/London, 1993), with Michael Inwood (*A Heidegger Dictionary*, Blackwell Publishers, Oxford, 1999), and others.

Kenneth Mills

In the editing process, all Greek terms were transliterated, the few French words were translated in brackets following the phrase, and we have provided some of the original German when the specific words used are worth noting.

In common with many recent translations we have translated *das Sein* as "being" and *das Seiende* as "a being" or "beings." In cases where there may be ambiguity we have provided the German. We have translated the phrase *das Nichts* as the Heideggerian "the nothing" rather than the more Sartrean "nothingness." *Dasein* is the only word left untranslated—its meaning is at once "existence" and a technical term in Heideggerian thought. In one instance it appears in a quote from Marx—there we have translated as "existence" and provided the German.

Axelos uses both parentheses and square brackets in the text; the latter especially to mark his interpolations in passages he quotes. These are usually clearly marked, and should not therefore be confused with translator additions.

Axelos provides both footnotes and some in text references. We have taken these all into the notes, where we additionally provide modern edition references (in the case of Heidegger and Nietzsche) and available English translations. We have not followed existing translations in all instances, to ensure consistency throughout the text. The page numbers are provided for reference and wider context. Two abbreviations are used:

– GA: Martin Heidegger, *Gesamtausgabe*, ed. Friedrich von Herrmann et. al. (Frankfurt am Main: Vittorio Klostermann, 1975).
– KSA: Friedrich Nietzsche, *Kritische Studienausgabe*, ed. Giorgio Colli and Mazzimo Montinari (Berlin and Munich: Deutscher Taschenbuch Verlag and Walter Gruyter, 1980).

If not marked, notes are by Axelos. Translator and editor notes are marked as TN and EN, respectively. In one instance where Axelos's comments follow an editor note, this is marked AN. I have provided a bibliography of Axelos's books, with indications of English translations in part or whole.

Stuart Elden

INTRODUCTION TO A FUTURE
WAY OF THOUGHT

KOSTAS AXELOS

Foreword

The following essays, lectures, theses, texts, and discussions, which were written, held, and repeatedly revised in the German language between 1956 and 1966, represent an attempt to introduce the thought of the future in a multi-dimensional yet holistic manner. They think along with Marx and Heidegger as well as about them. *About* [*Über*]—as dictionaries tell us—can mean not only *peri, de, von, sur*, but also *metá, trans, hinüber*, and *au-delà*.[1] By thinking through Marx and Heidegger, we can eventually think beyond them. At the same time, this form of reflection also engenders anticipatory thought. The thought of the future we must be introduced to is inseparable from thought which has gone before and from current thought—as well as from that which has not yet been thought. Moreover, this thinking should also be inseparable from past, current, and future experience. The experience of the world. Thus, the matter at hand is to allow a form of world thinking corresponding to worldly experience and

---

1 EN: The first set of words—Greek, Latin, German, French—mean "on" or "about"; the second set, "beyond".

its praxis to evolve, a kind of thinking which obeys both the prose and poetry of the world.

Such an elementary and future way of thought, together with a traditional, a becoming, and a failing lifestyle, belongs to the selfsame reality: the fragmentary one-all, which unfolds within time as *errance* and *the game*,[2] and then expires. In this way, reflection leads to the anticipatory thought into which we are initiated. The new reveals the old, which had concealed itself. This anticipatory thought is not noncommittal prophecy. The thought of the future—predicted and foreseen, predicting and foreseeing—corresponds to the *futurum* approaching us (which corresponds to the growing *physis*).

The reflective and anticipatory thought attempted here reaches back to the roots established by the poetic thought of Heraclitus, it positions itself beneath the constellation of Hegel, Marx, Nietzsche, and Heidegger, but also strives to supersede them: in order to introduce into the mundane game a new way of thinking fragmentary wholeness—at once worldly and historical, planetary, open, multi-dimensional, inquisitive, and playful. The mundane game outplays everything.

For today and tomorrow, world history appears to be dominated by a fusion of technology and science. This dominance itself unfolds in the form of what we call world history. The predominant technology places everything in question and must be questioned itself. Does *planetary technology* require *planetary thought*? Technicized science and scientific technology are taking over art, religion, and philosophy, and dissolving them. However, the fusion of technology and science does not solve the tasks of thought. Can this fusion evade thought and can thought evade this fusion? Is such a thing even possible? Might it lie within the potential of a coming—and already pre-existent—mode

2   TN: Italics from the translator. These are standard terms from the Axelos corpus.

of thought to experience its own power and powerlessness beyond idealism and materialism, beyond the metaphysical or anti-metaphysical dichotomy or gap, while intimately associated and merged with action? Can metaphysics be overcome and not merely reversed? This attempt places demands on our efforts. But how can the play of thought execute a transcendence that has somehow already achieved completion?

For some time now, humanity—as individuals, as members of society, and as fragments of world history—has been striving to live and to love, to speak and to think; humans work, they fight, they die. For what? For questionable pleasures and attainable prosperity—whereby more and more is never enough for them—, for recognition—which has of course become an increasingly relative notion—, for the realization of peace on earth and all-encompassing, negotiated reconciliation? What drives them on and where are they being driven to? And why? What can we still make of logical or mythological causes, or the power of historical momentum, or the aims pursued by humanity? Do they vacillate in their play? They materialize in a history that has already played out, they dissipate, they repeat and become outplayed once again without end. Does this mean the only recourse we have—without backward-looking desire, without a crude depreciation of the present, and without eschatological utopias—is to pursue the game with yet more width, depth, and breadth? Without a "why"?

The future way of thought is not only something that always remains a thing of the future. It was already there, it is already here, it has yet to arrive. It can demystify us and encourage us somewhat, but at the same time it will experience and effect new one-dimensionalities and superficialities. It is part of the new wealth and the new misery—which are and remain inseparable from the old way of living and loving, speaking and thinking, working and fighting, playing and dying. This is how it "is" with and in the play of time, which gathers and annihilates everything, while resting in its wholeness. This is the play of errance in whose game the foundation of all motion rests.

I: MARX AND HEIDEGGER

Marx and Heidegger: Guides to a Future Way of Thought

We are nothing; what we seek is everything.
(Hölderlin, preliminary notes to *Hyperion*[1])

I

In a letter to his Parisian friend Jean Beaufret, in the "Letter on 'Humanism,'" Heidegger writes: "Homelessness is coming to be the destiny of the world. Hence it is necessary to think of this destiny in terms of the history of being. The alienation of humanity, which Marx recognized in an essential and significant sense based on Hegel, has roots extending back to the homelessness of man in the modern era. This has been summoned, indeed, from the destiny of being in the form of metaphysics and concretized by the latter, but concealed by it as homelessness at the same

---

1   EN: Friedrich Hölderlin, "Fragment von Hyperion," in *Sämtliche Werke und Briefe in drei Bänden*, ed. Jochen Schmidt, vol. 2, *Hyperion. Empedokles. Aufsätze. Übersetzungen* (Frankfurt am Main: Deutscher Klassiker Verlag, 1994), 199.

44   time. As Marx has entered an essential historical dimension with
the experience of alienation, the Marxist [it would have been
better to state: Marxian][2] view of history is superior to competing
versions. However, due to the fact that neither Husserl nor—
as far as I have seen up until now—Sartre has recognized the
essential nature of historicity in being, neither Phenomenology
nor Existentialism is able to enter that dimension within which
a productive dialogue with Marxism first becomes possible."[3]
Thus, a productive dialogue with Marxian thought—and with
Marxism—should be sought here, and it could well be the case
that it is extremely important to facilitate such a dialogue. We
read further: "All the same, in this context it is also necessary
that one frees oneself of naive conceptions about materialism
as well as the cheap refutations meant to attack it. The essence
of materialism does not exist in the mere claim that everything
is material, but much rather in the metaphysical determination
which maintains that all beings [*alles Seiende*] appear as the
material of work. The metaphysical essence of work characteristic of the modern era is prefigured in Hegel's *Phenomenology
of Spirit* as the self-manifesting process of unconditional
production, it is the substantialization of the real through the
human individual experienced as subjectivity. The essence of
materialism conceals itself in the essence of technology, which is
indeed much discussed but seldom thought about." Heidegger
attempts to trace the materialism of Marxism and communism
back to the essence of work and technology. And he continues:
"In its essence, technology is a destiny of the history of being
whose truth lies in oblivion. In fact, it not only goes back to the
tékhne of the Greeks in name, it also originates from *tékhne* in the
history of being as a mode of *aletheuein*, which means a mode of

2   TN: The passages in brackets were added to the original text by Axelos.
3   Martin Heidegger, *Über den Humanismus* (1947; Frankfurt am Main: Vittorio
     Klostermann, 1949), 27. EN: "Brief über den Humanismus," in *Gesamtaus-
     gabe*, vol. 9, *Wegmarken*, ed. Friedrich-Wilhelm von Herrmann, 339; "Letter
     on 'Humanism,'" in *Pathmarks*, ed. William McNeill (Cambridge: Cambridge
     University Press, 1998), 258–59.

revelation of beings. As a form of truth, technology has its ground **45**
in the history of metaphysics. This itself is an excellent—and up
to now—the solely assessable phase in the history of being. One
can have varying opinions about communist doctrines and their
foundations, but in terms of the history of being it is evident that
an elementary experience of world history has been expressed
in them. Whoever takes 'communism' to be merely a 'party' or
'worldview [*Weltanschauung*]' is thinking as cursorily as those who
use the term 'Americanism' and mean only a lifestyle, and this in
a derogatory manner."[4]

In order to begin a discussion with Marxism, should we not first
involve ourselves in a dialogue with Marx himself? But with
which Marx? The philosopher? The economic scientist? The
politician? Does Marx have many faces, and what lies hidden
behind the various layers and faces? Perhaps Marx has ceased to

4   Heidegger, *Über den Humanismus*, 27–28. EN: GA, vol. 9, 340–41; "Letter
on 'Humanism,'" 259. AN: How different Jaspers's reflections on Marxism
sound. In one of the three guest lectures he held in Heidelberg (1950),
Jaspers intends to present to his listeners a concise, compact, and uniform
sketch of Marxian thought—as national economics, philosophy, and ethics,
and of Marxism—as "science" and "faith." The title of the lecture is: "The
Demands of Scientificity [*Forderung der Wissenschaftlichkeit*]." (These three
lectures also appeared in print: *Vernunft und Widervernunft in unserer Zeit*
(Munich: Piper, 1950) [EN: translated by S. Godman, as *Reason and Anti-
Reason in Our Times* (London: SCM Press, 1952)]). The lecture is intended
to address Marxism and psychoanalysis. The result is a presentation in
school philosophy, a brief summary of basic Marxist (and psychoanalytic)
principles, and a myopic interpretation of Marxism as a "substitute religion,"
as alleged absolute knowledge, which must be exposed: "The absolute
knowledge of Marx must be exposed as a form of the supposed knowledge
already realized by Hegel, which Marx reiterated in this form in an old-
fashioned manner but with specifically modern content" (14 [EN: *Reason and
Anti-Reason*, 16]). Admittedly, a particular form of truth is also attributed
to Marxism—professorially. In effect, however, the whole of this absolute
knowledge, this absolute plan in its entirety, succumbs to "fundamental
erring," the "wrong track," "error" and "going astray"—understood as "monstrous falsity"—, contrasted with those illuminated by truth which informs
"possible existence through reason." These are the reflections of Jaspers, a
proponent of "unlimited communication."

be a philosopher. In a lecture held in Cerisy (Normandy), "What is Philosophy?," Martin Heidegger poses the question: "Where should we seek the culmination of philosophy in the modern era? With Hegel or first in the later philosophy of Schelling? And what is the case with Marx and Nietzsche? Have they already abandoned the path of philosophy in the modern era? If not, how should their status be determined?"[5]

Aristotle, Thomas Aquinas, Hegel appear to us as the philosophers who have set standards and laid down the law. Behind and between them stand the pre-Socratics, Plato, Augustine, Descartes. All of them originated from an historical soil [*Boden*], which they influenced and which was influenced by them. Plato-Aristotle-Augustine-Thomas Aquinas-Descartes-Hegel dominate Greek-Christian-Modern thinking. Plato establishes Platonism, i.e., philosophy and any philosophy as idealism and dualism, Aristotle the onto-theo-logical methodology. Augustine Christianizes Platonism and Platonizes Christianity, Thomas Aquinas arms Christianity and allows it to speak with Aristotelianism. Descartes introduces the ingenious compromise of the modern era between the Platonic-Aristotelian idea, the Christian belief in revelation, and the "liberated" thinking ego of modern man—*res cogitans*, which is argued to dominate, possess, govern, and transform *res extensa*, within the confluence of philosophy and science. Somehow, Hegel concludes this entire chapter synthetically, historically, and systematically. All of these thinkers are interconnected in sundry ways, follow and pursue each other. From Plato-Aristotle, it runs—through the Roman era—to the Christian theology of Augustine and Thomas Aquinas, and from them—through Descartes and Kant—to Hegel. And

5   Martin Heidegger, *Was ist das – die Philosophie?* (Pfullingen: G. Neske, 1956), 42. EN: "Was ist das – die Philosophie?," in GA, vol. 11, *Identität und Differenz*, 24; *What is Philosophy?* trans. and introd. Jean T. Wilde and William Kluback, German-English edition (Lanham, MD: Rowman & Littlefield, 1956), 88–89. Axelos was the interpreter at this lecture, and the translator of the published French text.

from Hegel it leads to Marx and Marxism, which strive to overturn
the idealistic and speculative dialectic of Hegel both materialistically and practically. "The fact that medieval theologians study
Plato and Aristotle, i.e., reinterpreting them, is comparable to
how Karl Marx appropriates Hegelian metaphysics for his political
agenda. Viewed correctly, however, it is not the intention of
the *doctrina christiana* to provide knowledge about beings [*das
Seiende*] and what this actually means, as its truth is absolutely
a truth of salvation. Its objective is ensuring salvation for the
individual immortal soul." These are the somewhat uninspired
comments about the relationship between Marx and Hegel in
Heidegger's book about Nietzsche.[6] For all productive interpretations are reinterpretations, attempts to tame the Odyssean
voyage of world history—including the world history of thought.
Nostalgia and the yearning for faraway places correspond to one
another, like departure and return. Upheavals, reinterpretations,
revaluations belong with the very things they contest to the same
adventurous process, which works off opposites and opposing
forces both uniformly as well as alternately. Within this process,
Nietzsche appears as the one who strives to reverse and reval-
uate Platonism-Christianity; he himself considers his thought to
be "Platonism turned round"; thus, as overturned Christianity as
well, if Christianity is understood as "Platonism for the 'people.'"
At the same time, Nietzsche intends to enter the fray as the
identifier and over-comer of nihilism.

As a result, Platonism-Aristotelianism, Christianity, Cartesianism,
Hegelianism-Marxism-Nihilism—their entire development and
turning points, which include all upheavals—are and remain the
dominant course of a reeling world history. Whether this course
is experienced in the pre-philosophical mode (with Heraclitus
and the pre-Socratics), in the philosophical-metaphysical mode,
i.e., onto-theologically (from Plato to Hegel), in the devout and

6   Martin Heidegger, *Nietzsche*, vol. 2 (Pfullingen: G. Neske, 1961), 132. EN: GA,
     vol. 6.2, *Nietzsche II*, 116; *Nietzsche*, vol. 4, *Nihilism*, ed. David Farrel Krell,
     trans. Frank A. Capuzzi (San Francisco: Harper & Row, 1982), 88.

Christian mode, in the bourgeois mode, socialist mode, nihilistic mode, or even anti-philosophical and anti-metaphysical modes, as well as non-philosophical and extra-metaphysical modes, it must be ascertained, or more importantly—experienced, whether the meta-philosophical path (opened by Marx and Nietzsche) can still be traversed.

Marx might well not be a philosopher any more, if philosophy begins with Plato—as philosophy—and achieves its completion with Hegel, which must not imply that completion, demise, and conclusion have converged here, let alone that their *telos* remains puzzling. Marx might well not be a philosopher any more, but still belongs to the modern era, to the epoch of subjectivity; he could even be one who wishes to generalize subjectivity and raise it to the level of a societal category. Nonetheless, even as a non-philosopher, perhaps even as a wayfarer, Marx remains a thinker. What does he think? In the "Letter on Humanism," Marx's thought and main direction are characterized briefly and accurately as follows: "Marx demands that the 'human's humanity [*menschliche Mensch*]' be recognized and acknowledged. Marx finds it in 'society.' He views the 'social human' as the 'natural human.' In 'society,' the human 'nature,' i.e., the whole of 'natural needs' (nourishment, clothing, procreation, economic survival) [the elementary?] is equably secured."[7] Heidegger places the terms "human's humanity", "society," "nature," and "natural" in quotation marks. Why? Because Marx makes use of them? Not only, and not primarily, for this reason. Rather, because they are problematical terms, since we do not quite know how we should comprehend them. The demands placed by Marx—the resolution of externalization and divestiture, the radical resolution of alienation through the destruction of its "real" foundations, of private property, of the capitalistic division of labor as well as the traditional means of production and work, and the

7   Heidegger, *Über den Humanismus*, 10. EN: GA, vol. 9, 319; "Letter on 'Humanism,'" 244.

acknowledgement of the natural, societal, human's humanity— are and remain problematical. But they nevertheless remain demands which allow elementary forces to express themselves and which put these forces to practical application.

Does Heidegger himself attempt to enter into a dialogue with Marx? Does he initiate a productive discussion with Marxism? Heidegger does not often mention Marx. In spite of this, his own thinking has ties to Marxian thought. In the year 1932, Marx's most important philosophical work was published, the Parisian economic-philosophical manuscripts of 1844, "Nationalökonomie und Philosophie" [National Economy and Philosophy];[8] this work had not appeared in print prior to this. The first 1932 edition, with Kröner, was obtained by Siegfried Landshut, who was a student of Heidegger, and in this way Heidegger was granted direct access to the Parisian manuscripts. Since this point in time, this profound but also incomplete and fragmentary text has required a correspondingly profound interpretation, which in turn presupposes an interpretation of Hegel, as the main topic on this Marxian work is a coming to terms with Hegel, and most pointedly, with the *Phenomenology of Spirit*, "the true birthplace and secret of Hegelian philosophy" according to Marx's own formulation.[9]

Heidegger does not provide us with the guidelines of a Marx interpretation. He never occupied himself with Marx to any significant extent, as was the case with Anaximander, Heraclitus, Parmenides, Plato and Aristotle, Leibniz, Kant, Hegel, and Nietzsche. In spite of this, Marx was not absent from Heidegger's intellectual ventures by any means. By attempting to grasp the essence of modern technology—even of planetary technology—,

---

8   EN: This is a literal translation of the title Axelos uses. The more standard
     English title is simply "Economic and Philosophical Manuscripts" or "1844
     Manuscripts." A complete translation under the first of these titles is found
     in *Early Writings*, trans. Rodney Livingstone and Gregory Benton (London:
     Penguin in association with New Left Review, 1975), 279–400.
9   EN: Marx, "Economic and Philosophical Manuscripts," 382–83.

by attempting to bring light to the roots of the era of mechanization—and of the atomic age—, by attempting to intellectually question the world destiny of homelessness and rootlessness, Heidegger gives us the sense that Marx stands behind this endeavor[10] The object of this questioning is the essence of

10 There are Parisian Marxists, e.g., Lucien Goldmann and Joseph Gabel, who are of the opinion that a relationship exists between Heidegger and [Georg] Lukács. First of all, they want to draw attention to a number of biographical facts: During the pre-WWI period, Heidegger and Lukács moved in the wake of the Southwestern School of Neo-Kantianism. Both stood in a close relationship with Emil Lask. Secondly, and primarily, this opinion has the objective of substantiating that Heidegger had adopted important topics from Lukács. *Die Seele und die Formen* and *Geschichte und Klassenbewusstsein* by Lukács appeared in 1911 and 1923. *Sein und Zeit* appeared in 1927. Themes taken up by Lukács which constitute the tragedy of life, reification, totality, history, and conscious being, are argued to have become Heidegger's inauthentic and authentic Dasein, being-toward-death, objectification, being as a whole, time, historicity, and being. Both strive to resolve the reification and objectification predominant in the capitalist and modern world with a new historical concept of being, and to open a new historical horizon. The comparison between Lukács and Heidegger is intended to underscore a predominance by Lukács. Accordingly, the young Lukács is purported to have been a catalyst for Heidegger. A history of influences and encounters, anticipations, etc., always remains a confusing affair. The tendency implied by this comparison is quite evident. Does this reveal anything of significance? Does the Marxist, historical and philosophical, ideological and sociological school of thought espoused by Lukács exist at the same level as the metaphysical and ontological school of thought established by Heidegger with its failure to think the history of being, which is intended to overcome philosophy (metaphysics)? Should we not seek the roots of Heidegger's thought in a different place and with different means? When such questions are asked, this simultaneously identifies the locus of their suitable answer. —The young Lukács grew older, and as a result, he entered another dimension: within the latter, he complained about Heidegger and piecemealed what he was no longer able or willing to comprehend sufficiently with polemical intent: see "Heidegger redivivus" (in *Sinn und Form* 1949, third no., a polemical work against the "Letter on 'Humanism;'" *Existentialismus oder Marxismus?* (Berlin 1951); *Die Zerstörung der Vernunft* (Berlin 1954). EN: The texts by Lukács mentioned by Axelos are, respectively: *Die Seele und die Formen* (Berlin: Egon Fleischel & Co, 1911); translated by Anna Bostock as *Soul and Forms*, ed. John T. Saunders and Katie Terezakis (New York: Columbia University Press, 2010); *Geschichte und Klassenbewußtsein:*

technology. Industrial labor and machines are manifestations of technology whose essence is still concealed from us. Marx enters an essential historical dimension; despite this, he could not yet fully answer the fundamental questions concerning technology, as we today are also still unable to. In his Freiburg lectures, "What is Called Thinking?" (1951/52), one could hear and now one can read the following: "Neither the industrial laborer, nor the engineers, nor even the factory owners, and least of all the state, are able to conceive of how humanity is actually situated today, when it exists in any relation to machinery or parts of a machine. None of us are already aware about the kind of hand-work modern man must undertake in the technological world, and still must undertake even when they are not a worker in the sense of a laborer at a machine. Neither Hegel nor Marx were able to know this and ask the right questions, as their thinking could not yet avoid being obscured by the shadow of the essence of technology, for which reason they were never able to reach the free space where they could perceive this essence sufficiently."[11]

Heidegger has often been accused of animosity towards technology, the rejection of technology, as well as a regressive if not reactionary attitude and approach. According to this all too widespread critique, romantic desire as well as an unmodern and non-progressive attitude characterize Heidegger's work, to which the accusation of its unscientific quality is often added. Are these claims substantiated? Even if something correct is expressed in this manner, does this critique manage to reach the level of the essential? Does it address the fundamental tone and main

Studien über marxistische Dialektik (Berlin: Malik, 1923); translated by Rodney Livingstone as *History and Class Consciousness* (London: Merlin, 1971); "Heidegger Redivivus," *Sinn und Form*, no. 3 (1949): 37–62 (untranslated); *Existentialismus oder Marxismus?* (Berlin: Aufbau, 1951) (untranslated); *Die Zerstörung der Vernunft* (Berlin: Aufbau 1954); *The Destruction of Reason*, trans. Peter R. Palmer (London: Merlin, 1980).

11   Martin Heidegger, *Was heißt Denken?* (Tübingen: Niemeyer, 1954), 54–55. EN: GA, vol. 8, *Was heißt Denken?*, 27; *What is Called Thinking?*, trans. J. Glenn Gray and Fred D. Wieck (New York: Harper & Row, 1968), 24.

direction of Heidegger's intellectual venture, which, as he says himself, is an *attempt* as well as a way of thinking which acquires its own failure as a gift? Is it indeed the case that this *failing thought* does not want to take note of the dawning planetary age? Must an age directed at success remain closed and foreign to such a failing intellectual venture, or are the crisis itself, the breach, and the alienation, which inhabit this epoch, coming up for discussion here?

Heidegger warns his readers about the passage quoted above regarding Hegel and Marx, whose thinking could not yet avoid being obscured by the shadow of the essence of technology. He warns about the kind of misunderstanding that rashly reinterprets that which has been said and heard. When Heidegger speaks, he frequently speaks of shepherds, farmers, and land, of field paths and timber tracks [*Holzwege*], of trees and mountains. All of that sounds agrarian, farmer-like. The Black Forest seems to be a dark mountain that obscures the lucid view of the unfolding of modernity. In the lecture *What is Called Thinking*, where he attempts to speak of the essence of technology and thinking as exceptional handwork, he chooses as an example the craftsmanship of the village carpenter. He chooses *this* example. Certainly. But he warns: "The craft of the carpenter was chosen as an example and at the same time it was assumed that nobody would fall for the idea that choosing this example amounts to the announcement of the expectation that the condition of our planet can be transformed into a rural idyll in the near future, or indeed ever again."[12] Thus, the matter at hand is not to expect that the condition of the earth will ever be transformed back into a rural paradise.

Marx speaks very explicitly about the externalization, divestiture, objectification, alienation of modern man. Heidegger speaks of the objectification of all things which have being through

---

12   Heidegger, *Was heißt Denken?*, 53–54. EN: GA, vol. 8, 26; *What is Called Thinking?*, 23.

the will of subjectivity, of the homelessness of humanity in the modern era, of the abandonment by being, of the oblivion of being. Marx says that technicized labor leads to the denial of the human, and not only humans are alienated from their essence, but objects from their essence as well. Concerning a human he states: "as their life expression is also an alienation of life, their realization is an undoing of reality, an *alien* reality."[13] The realities undergoing alienation are the essential powers of the human, "the ontological essence of human passion,"[14] ". . . as all human activity up to now has been labor, i.e.: industry, an self-alienated activity—we have in the mode of alienation the *objectified essential powers* of humanity before us in the form of *sensual, foreign, useful objects*."[15] Marx's basic mood is not one of desire; Marx does not demand any return—"the humanism of Marx does not require a return to antiquity," is what we read in the "Letter on 'Humanism'"[16] —, he expects even less that world history reverts to a rural idyll. In spite of this, however,—or for this very reason—Marx can write: "The dwelling in the *light* [the word 'light' is emphasized by Marx], which Prometheus names as one of the great gifts with which he made humans from savages, ceases to exist for the worker."[17] All humans have become workers, and what Marx sees and says not only concerns industrial laborers: "The savage in his cave—that natural environment offering use and protection without reservation—does not feel more estranged, but feels as much at home as a *fish* in water. But the basement dwelling of the poor man is a hostile dwelling,

---

13   Karl Marx, "Nationalökonomie und Philosophie," in *Die Frühschriften*, ed. Siegfried Landshut (Stuttgart: Kröner, 1953), 239. EN: "Economic and Philosophical Manuscripts," 351.

14   Marx, "Nationalökonomie und Philosophie," 296. EN: "Economic and Philosophical Manuscripts," 375.

15   Marx, "Nationalökonomie und Philosophie," 244. EN: "Economic and Philosophical Manuscripts," 354.

16   Heidegger, *Über den Humanismus*, 11. EN: GA, vol. 9, 321; "Letter on 'Humanism,'" 245.

17   Marx, "Nationalökonomie und Philosophie," 256. EN: "Economic and Philosophical Manuscripts," 359.

which binds him like some alien power and which only responds to his blood and sweat, and which he cannot consider a refuge where he could finally say: I am at home here, for he exists in the house of another, in an *alien* house . . ."[18] After they are thrust into objectification and reification, the things themselves lose their essence: "Private property not only alienates the individuality of humans, but also that of objects as well. Land has nothing to do with ground rent, machines have nothing to do with profit." This is the way the founder of Marxism chose his words.[19]

And what does Heidegger say? He says: "The decline [of the truth of beings] occurs most evidently in the collapse of a world defined by metaphysics and the desolation of the earth caused by metaphysics. Collapse and desolation achieve their characteristic consummation based on the fact that the human of metaphysics, the animal rationale, has been 'held fast' [*'fest-gestellt'*] as a laboring animal."[20] Heidegger says: Work "achieves the metaphysical rank of the unconditional objectification of all that is present, which has become manifested as will to will."[21]

II

The passages quoted here from the writings of Marx and Heidegger are not simple quotations intended to prove something. They do not intend to prove—and they do not *intend* to prove *anything*—that Marx and Heidegger are naming similar things. At best, they can indicate what Marx and Heidegger—in

18   Marx, "Nationalökonomie und Philosophie," 266. EN: "Economic and Philosophical Manuscripts," 366.

19   Karl Marx, *Die deutsche Ideologie* (1845/46; Berlin: Dietz, 1953), 234. EN: *The German Ideology* (Moscow: Progress Publishers, 1964), 248.

20   Martin Heidegger, "Überwindung der Metaphysik," in *Vorträge und Aufsätze* (Pfullingen: G. Neske, 1954), 72. EN: GA, vol. 7, *Vorträge und Aufsätze*, 70; "Overcoming Metaphysics," in *The End of Philosophy*, trans. Joan Stambaugh (London: Souvenir Press, 1975), 86.

21   Heidegger, "Überwindung der Metaphysik," 72. EN: GA, vol. 7, 70; "Overcoming Metaphysics," 85.

differing ways—attempt to think of as the same thing: within
the cohesion of disjunction. The objection one can raise against
this juxtaposition is correct. The objection goes: Marx thinks
"ontically," and in accordance with logic and dialectic, which
means historically and epochally, sociologically, economically,
anthropologically, politically; he is concerned with experience,
true knowledge and the recognition of the empirical, real, realist,
practical, concrete, objective, and sensually perceptible states
of the modern world, in order to change it in an empirical, real,
realist, practical, concrete, objective, sensual and meaningful
way, through the true acknowledgement of the natural and social
human, which should then lead to the actual satisfaction of their
basic needs. Heidegger, on the other hand, thinks "ontologically
and metaphysically," as well as speculatively, which means in
terms of the history of being and the world, he questions the
difference, the rift between being [*Sein*] and beings [*Seiendes*]; he
is concerned with thinking being itself, the being lying dormant in
the oblivion of being, in a world that has become a surrounding
world [*Umwelt* / environment], and he attempts to utilize his questioning to overcome metaphysics as the history of this oblivion of
being, in order to open a new horizon in the receding clearing of
the world.

All of that is correct. Marx and Heidegger do not say similar things
by any means—their thought moves within the "selfsame." This
sameness encompasses both. Both stand on the bottomless
foundation of the modern era, the epoch of subjectivity; both
dare the attempt to overcome philosophy; both fight for a new
understanding of being. Marx does not construe the difference
between being and beings [*Sein und Seiendem*]; and perhaps
Heidegger himself does not focus on certain vital aspects of
being.

In a worldless world, one could reply in response to the
comparison between Marx and Heidegger that neither Marx
nor Heidegger are deep and significant thinkers who are able to
say anything about the world. But in the name of what thought

could the thinking of both of these thinkers be appraised? In the end, one could also reach a two-fold conclusion: that Marx is the greater of the two, subsuming Heidegger in his thought, so that only a Marxian—or even Marxist—interpretation of Heidegger remains. That Heidegger's thought becomes lucid and comprehensible in the light of Marxian thought. Conversely, one could be of the opinion that Heidegger subsumes Marx, that Marx—and Marxism—can be completely and satisfactorily subordinated to Heidegger's thought, or put roughly, that Heidegger supersedes Marx by going farther and taking a higher look. The history of philosophy influenced by doxography, philosophical studies, school philosophy, cultural philosophy, and cultural sociology tend to proceed in this way. They are not concerned with the "truth" but only with "opinions" that can be assimilated by one-dimensional thinking. The fact that a great thinker cannot supersede another great thinker, that the truth of thought and thinking the truth are and remain multi-dimensional, that the openness of the world cannot be exhaustively comprehended by one form of thought—as great as it may be, these are realizations denied by historians, ideologues, orthodox or heterodox supporters or opponents of "-isms," professors, and journalists. Higher-level ambiguity is none of their affair.

One wants to know, i.e., think very little about the subject of thought and being; one desires scientific interpretations above all. Thought and being? Being and thought? In the horizon of time? "Thought and being are indeed *different* [emphasized by the author himself] but simultaneously in *unison* with one another." Who pronounced this sentence? Marx or Heidegger? Is it Marxian or Heideggerian? Does it exist in Marx's Parisian manuscript or in Heidegger's letter to his Parisian friend?

In order to understand Marx and Heidegger really and truly, it would be necessary to conceive of the history of philosophy both metaphysically and metaphilosophically. The pre-Socratics loom in the background, and both Marx and Heidegger make reference to Greek thinkers of early antiquity. In the opinion of the one, Plato and Aristotle represent the beginning of systematic ideological alienation, whereas the other holds them to be the philosophical and metaphysical start of the oblivion of being. Christianity with its two worlds, as well as the claim to power of Christianity, are taken as seriously by the one as the other—but it must always be repeatedly asked in a banal tone: in different ways? The philosophy of German idealism and its culmination with Hegel, the last philosopher, are the points of departure for Marx. With Hegel, a great epoch ends and a complicated history begins. The collapse of German Idealism and Romanticism allows the rift in the world to appear. "The diremption [*Diremption*] of the world is not causal, if its sides are totalities. Thus, the world is ruptured and comes into relation to a philosophy that is an intrinsic whole. As a result, the appearance of the activity of this philosophy is also fragmented and contradictory; its objective generality converts to the subjective forms of individual consciousness where it is alive. Common harps can be played by any hand; Aeolian harps only respond to the force of a storm. However, one should not be led astray by this storm, if one is following a great, a world philosophy," are the words in the doctoral dissertation of the young Karl Marx, "The Difference Between the Democritean and Epicurean Philosophy of Nature."[22] But does this imply that after Hegel a new way of thinking has stormed into existence? As was mentioned above, an adequate interpretation of Marx presupposes an interpretation of Hegel. We do not have one yet. It is not at all as simple as it might seem to discuss the relationship between Marx and Hegel. Marx does not place the Hegelian dialectic on its feet in a crystal-clear manner. And what does he do with the head of the Hegelian dialectic? Does Marx merely transcend Kantian and Hegelian analysis? Does he in fact fall short of their standards as well? The

22 Karl Marx, "Aus der Doktordissertation" (1840), in *Die Frühschriften* 13. EN: "Notes to the doctoral Dissertation" (1839–41), in *Writings of the Young Marx on Philosophy and Society*, ed. Loyd D. Easton and Kurt H. Guddat (New York: Doubleday, 1967), 52–53.

fundamental approach taken by Marx can be used as a pivotal point: he no longer wishes to think speculatively and metaphysically, as this way of thinking remains subjected to alienation; he desires to ascertain in a practical mode and effect change in a practical mode. But what occurs during this transformation?

An immediate historical leap does not lead from Hegel and Marx to Heidegger. Kierkegaard and Nietzsche also appear on the scene. For his part, Kierkegaard considers himself to be an opponent of Hegel; he fights for the exception, for existence, for the (lost) connection to God; is he merely or primarily a "religious author," as Heidegger once described him? Nietzsche is a persevering puzzle: the lightning bolt that decimates everything, including the one who perceived it. Nietzsche's thought, his terminology and thoughts remain puzzling; the words "puzzle" and "puzzling" here are not intended to conjure up any secret: they are only an expression of what cannot yet be said. For what should we, may we, and can we say? God is dead. The will to power rules. Nihilism begins its reign in the growing desert. The *Übermensch* has not yet arrived. Nobody experiences the innocence of becoming any more. The eternal recurrence of the same hides itself and becomes blocked. The supernatural, metaphysical and theological, divine Truth that bestows all meaning, God as a figurehead [*Gestalt*] of divinity, the whole of godliness are no longer present. God has been murdered. By the human? By the human as the plenipotentiary of the will to power? Is the time of nihilism also the time-space of the world game that peruses and plunges the earth-globe [*Erdball*]—this wandering star [*Irrstern*]—into errant being [*ins Irrende wandernd*]? Hidden in a nihilism beyond good and evil, neither meaningful nor meaningless, can we find the erring "truth" itself, "being" thrust into becoming, perhaps even the "divine madman," whom Socrates was able to allude to so playfully? Let us reflect upon the Heraclitean-Platonic *Cratylus* where *truth, aletheia*, is interpreted as *ale theia, divine madness*, but only for a moment—in a Socratic and ironic, Heraclitean but Sophistic, misleading but truthful manner.[23] Does the nothing, which is itself destroyed by nihilism, annihilate being and its truth, annihilate beings and its meaning? Does nihilism open a new temporal and planetary play-room [*Spiel-Raum*], which is only lacking the *Übermensch*? Salvation from revenge, consummated nihilism and nihilism overcome, resolved detachment and indifference—will they become a world, an innocent becoming, and allow it to flourish? A becoming which no longer opposes being? As regards the eternal recurrence of the same, we can indeed see what it obscures: the rotational movement of technology, the orbit of the planet earth, the planetary rings, and the revolutions of world history. In all of these circles, the eternal recurrence of the same remains hidden, and eternity forms the mask of time.

Why have we attempted to repeat these questions from Nietzsche's thought? To what extent do they concern—or not concern—Marx and Heidegger? Hegel, Marx, Nietzsche, Heidegger . . . Four names, four thinkers; four thinkers, who—each at his own pace—take and fall off the same path [*auf demselben Wege gehen und fallen*].

III

Marx describes a "world order" of divestiture, externalization, foreignness, and alienation. Nevertheless, a new destiny can ignite the world and allow world history to first evolve through this event. The necessary possibility is: the resolution of alienation, a "complete and conscious return of the human on the basis of the entire wealth of the prior development"[24], a "true dissolution of the conflict between existence and essence, between objectification and self-affirmation, between freedom and

23 Plato, *Cratylus* 421b. FN: Plato, *Cratylus*, trans. Harold N. Fowler, Greek-English edition (Cambridge, MA: Harvard University Press, 1921), 421b.
24 Marx, "Nationalökonomie und Philosophie," 235. EN: "Economic and Philosophical Manuscripts," 348.

necessity."[25] This return to the essence of the human and world history, which has never been realized before, is a leap in the direction of the future and means the unfettering of technology and productivity. The "resurrection of nature," which Marx speaks of, is at the same time and predominately the "consummated humanism of nature," the practical acquisition of the natural being of the societal human. Marx also realizes on occasion that the essence of communism cannot manifest itself, that no *true* dissolution of the conflict must occur. He speaks of the "prerequisite" of socialism-communism, and even writes on a tattered and barely legible page of the Parisian manuscript: "When we still describe communism—as the negation of negation, as the acquisition of human essences which actualizes itself through the negation of private property—for these very reasons as not yet the *true* position [emphasis by Marx], as it does not begin with itself, but much rather with private property,"[26] this must imply that alienation has not actually been resolved. In spite of this, the path of abolishing private property must be taken. This measure must be carried out. "For us, communism is not a *condition* that should be established or an *ideal* according to which reality should direct itself. We call communism the *actual* movement that offsets the status quo. The conditions of this movement arise from the currently existing prerequisite," are passages from another work which also was not published by Marx himself.[27] The activity that establishes communism is not a final one: communism is "the *actual* momentum of human emancipation and self-recovery necessary for the coming historical development. *Communism* is the necessary form and energetic principle of the near future, but communism is not as such the objective

25   Marx, "Nationalökonomie und Philosophie," 235. EN: "Economic and
       Philosophical Manuscripts," 348.
26   Marx, "Nationalökonomie und Philosophie," 264. EN: "Economic and
       Philosophical Manuscripts," 365.
27   Marx, *Die deutsche Ideologie*, 32. EN: *The German Ideology*, 47.

of human development—the form of human society."[28] Communism—provided that it has manifested itself in accordance with its essence or even: according to its empirical manifestation, i.e., its "impure" and rough realization—will be overcome itself. The activities and the movement, the active process that lead to communism, in order to realize it, will be overridden. "History will cause it to happen [this active process] and that movement, which we already know *in thought* as one that supersedes itself [*eine sich selbst aufhebende*], will undergo a very crude and wide-ranging process in the empirical world. Nonetheless, we must consider it to be true progress that we have acquired in advance a consciousness that both apprehends limitation as the objective of historical development and is able to transcend it as well."[29] Thus spoke the founder of Marxism, prior to the actualization of communism, about the negated process of socialization; what has already become negated in thought can only be overcome and negated within the process of its concrete manifestation (which is simultaneously negated manifestation). Not only the young Marx thinks this way. The young Marx cannot be separated from the mature and old Marx any more than the early Heidegger from the later, although there is a bridge that leads from their early to their later thought. The thought of a mature thinker can be even more elementary and more original, or also guide one onto steadier tracks—in relation to its initial development. In any case, a Marx who has grown older writes in the "Foreword" to the first volume of *Capital* (1867) what the younger Marx had already said earlier: "Along with modern exigencies, we are burdened by an entire series of inherited exigencies which have arisen from the continuing vegetation of ancient, outmoded means of production with their train of *anachronistic* social and political conditions. We do not only suffer because of the living, we also suffer because

---

28 Marx, "Nationalökonomie und Philosophie," 248. EN: "Economic and Philosophical Manuscripts," 358.

29 Marx, "Nationalökonomie und Philosophie," 265. EN: "Economic and Philosophical Manuscripts," 365.

of the dead. *Le mort saisit le vif!*"[30] The dead (*le mort*) which assails the living (*le vif*) does not possess merely an economic, political, and social character.

The world order intended to bring order to the world as a world, does not appear to be a matter for tomorrow. Perhaps it is not even possible to bring the world in order with order. "Humanity strives in vain to bring the earth-globe in order with its plans," are the words of Heidegger wandering along his field path [*Feldweg*],[31] where he dares to go off the beaten track [*Holzwege*] "World-withdrawal and world-decay cannot ever be reversed," he writes in *Off the Beaten Track*.[32] So—what should we think and do? Go marching ahead? To where? For what? Prepare new growth and cultivate it immediately? "As reality consists in the regularity of predictable calculation, the human must also cope with uniformity in order to measure up to reality"; such is the message of a phrase from that intellectual pursuit dedicated to "overcoming metaphysics," i.e., the overcoming of philosophy as "correspondence that gives speech to the beckoning of the being of beings"[33],—a correspondence which allows the truth of being to rest within the oblivion of being. Marx—although in a different sense—also aimed for the negation of philosophy, to which Nietzsche likewise made a powerful contribution. Is it all

30   Karl Marx, *Das Kapital: Kritik der politischen Ökonomie*, Bd. 1 (Berlin: Dietz, 1955), 7. EN: "Preface to the First Edition," in *Capital: A Critique of Political Economy*, vol. 1, trans. Ben Fowkes (London: Penguin in association with New Left Review, 1976), 91.

31   Martin Heidegger, *Der Feldweg* (Frankfurt am Main: Vittorio Klausmann, 1953), 4. EN: "Der Feldweg," in GA, vol. 13, *Aus der Erfahrung des Denkens 1910-1976*, 89; "The Pathway," trans. Thomas F. O'Meara, rev. Thomas Sheehan, in *Heidegger: The Man and the Thinker*, ed. Thomas Sheehan (New Brunswick, NY: Transaction Publishers, 1981), 69–71.

32   Martin Heidegger, "Der Ursprung des Kunstwerkes," in *Holzwege* (Frankfurt am Main: Vittorio Klostermann, 1950), 30. EN: GA, vol. 5, *Holzwege*, 26; "The Origin of the Work of Art," in *Off the Beaten Track*, ed. and trans. Julian Young and Kenneth Haynes (Cambridge: Cambridge University Press), 20.

33   Heidegger, *Was ist das - die Philosophie?*, 46. EN: GA, vol. 11, 26; *What is Philosophy?*, 96–97.

that important to stress the differences here? Do differences still exist? The oblivion of being forgets the divide [*Unterschied*]—the difference [*Differenz*]—, which holds sway between being [*Sein*] and beings [*Seiendes*]. But has not everything today become undifferentiated to the highest degree, indifferent, detached? "Indiscrimination affirms the priorly ensured existence of the non-world of the abandonment by being. Earth appears as the un-world of errancy [*Irrnis*]. In terms of the history of be-ing [*Seyn*], it is the wandering star [*Irrstern*]".[34] Errance itself, which is more deeply fundamental than going astray, nevertheless demands to be experienced and apprehended by thought. Philosophy, as metaphysics, was unable to think errance in a truthful manner; can then metaphysics, which culminates in planetary technology, still provide assistance in understanding errance? But if a triumphant metaphysics was unable to accomplish this, how can a philosophy that has already attained completion successfully analyze the "essence" of errance? Will a future way of thinking be capable of experiencing the "truth" as "errance" and "errance" as the truth—but without any disjunction whatsoever? It appears that it would be necessary first of all to consider being on the wrong track, the destiny of the wandering star in world history. The planetary form of thought, which has been given this dangerous gift, still lingers behind the planetary thought of the future. The planetary horizon has not yet opened. Let us return once again to Heidegger's intellectual attempt to *overcome metaphysics* in order to be able to dare later to advance even further; one reads: "Metaphysics in its completion, which is the ground for the planetary form of thought, provides the structure for a presumably long-term order of the earth. This order does not require philosophy any more, as philosophy already forms its basis. But the end of philosophy does not imply that thought has also reached its end, it is much rather in

---

34   Heidegger, "Überwindung der Metaphysik," 97. EN: GA, vol. 7, 96; "Overcoming Metaphysics," 108–9. Seyn is Heidegger's archaic spelling of Sein—beyng in place of being, or be-ing.

transition to another beginning."[35] The planetary form of thought certainly does not imply the consummation of planetary thought exclusively. On the entire surface of the earth, the planetary form of thought generalizes the whole history of metaphysics and its mode of thought scientifically and philosophically, ideologically and metaphysically, anthropologically and psychologically, in a universally historical manner and sociologically, literarily, and aesthetically. The planetary form of thought necessarily belongs to planetary technology, and the union of both is bound to the becoming of the wandering star. As a consequence, *planetary thought*, provided that it may be termed as such, experiences errance as the history of being and the world, espies as an intellectual faculty a new beginning, and is directed towards the future. By attempting to liberate world history from the burden of the brute past in order to allow time to temporalize and manifest itself as the having-been, as presence, and primarily, as the future, Marx and Heidegger are perhaps guides to this future thought of tomorrow. The "destruction" which they carry out fragmentarily is the destruction of the externalized production of life and its ideological superstructure, the destruction of an understanding of being still bound by metaphysics and lost in the oblivion of being.

Both know that world and the human (being and Dasein)—as they are—are at the same time nothing; what they seek is everything—the one-all. Naturally, i.e., historically, both of them are also in error. But perhaps not enough so. *The Thinker as Poet* gives us a hint, as Heidegger writes: "Whoever thinks greatly, must err greatly"[36]. This erring, however, should not be misunderstood in an all too erroneous and trite manner. Marx and Heidegger are on the way to turning logic and rationality upside

---

35   Heidegger, "Überwindung der Metaphysik," 83. EN: GA, vol. 7, 81; "Overcoming Metaphysics," 95–96.

36   Martin Heidegger, *Aus der Erfahrung des Denkens* (Pfullingen: G. Neske, 1954), 17. EN: "Aus der Erfahrung des Denkens," in GA, vol. 13, 81; "The Thinker as Poet," in *Poetry, Language, Thought* (New York: Harper Perennial, 2001), 9.

down. It is not at all the case that Marx is attempting to grasp the transformation and the "conversion of history into world history"[37] in the name of "reason." He is not even striving for a dialectical logic, let alone that he had ever spoken of "dialectical materialism." But just as little as Heidegger does he demand anti-reason, irrationalism. When Heidegger dares to say: "Thinking first begins when we have experienced that reason, which has been glorified for centuries, is the most stubborn nemesis of thought," he is not merely wandering *Off the Beaten Track*;[38] he is taking intellectual pains, in a *preliminary* manner, to prepare the way for the planetary thought of the future, for a language that errs and thinks, and attempts to develop a questioning reply to the beckoning of planetary destiny in world history, but which perseveres in openness without ever becoming inflexible or rigid.[39]

IV

Who among us does not know that Marx is concerned with praxis, with "*practical*, human and sensual activity,"[40] and that the main point is to *change* the world concretely and revolutionarily

37 Marx, *Die deutsche Ideologie*, 44. EN: *The German Ideology*, 48.
38 Martin Heidegger, "Nietzsches Wort 'Gott ist tot,'" in *Holzwege*, 247. EN: GA, vol. 5, 267; "Nietzsche's Word: 'God is Dead,'" in *Off the Beaten Track*, 199.
39 As a convention under Heidegger's supervision was taking place in Cerisy-la-Salle (Normandy) in the August of 1955, and after the lecture "What is Philosophy?" / "Was ist das – die Philosophie?" (*"Qu'est-ce que la philosophie?"*) had been held, a turbulent and confused discussion broke out. Everyone wanted to finally know the truth about Heidegger's thought. At the end of the initial day of discussion, Heidegger ended the session with the following words of the painter Braque: *Les preuves fatiguent la vérité* [EN: Proofs tire truth]. And after seven days, at the conclusion of the session, Heidegger once again closed with another excellent thought from Braque: *Penser et raisonner font deux* [EN: Thought and rationality are two things]. Another notation from Braque indicates the same; it reads: *L'erreur n'est pas le contraire de la vérité* [EN: Error is not the opposite of truth].
40 Karl Marx, "Thesen über Feuerbach," no. 8. EN: This is actually a reference to Thesis no 5. "Concerning Feuerbach," in *Early Writings*, 422.

instead of merely *interpreting* it in various ways philosophically and theoretically?[41] Everything, the entirety of beings, should enter the circle of productive human praxis and be brought forth through the same; certainly: this "everything" should be reflected by "human praxis" and gain its form through the latter; however, Marx also adds: "and by the comprehension of this praxis"[42]; thus, the intellectual comprehension of intervening praxis appears on the scene, although as a subsidiary power and machination. On the other hand—but what does "on the other hand" mean when Heidegger himself calls our attentions to it: "If there were adversaries in thought and not merely opponents, the situation for thought would be more favorable"[43]—isn't it the same Heidegger who demands that we liberate ourselves from a technical interpretation of thought? "Its beginnings date back to Plato and Aristotle. Thought itself is regarded as *techne* there, a process of deliberation in the services of doing and making. In this context, however, deliberation is viewed in relation to *praxis* and *poíesis*. For this reason, thought—regarded on its own terms—is not 'practical.' The characterization of thought as *theoria* and the determination of apprehension as a 'theoretical' behavior occurs within the 'technical' interpretation of thought. It represents the reactive attempt to save the autonomy of thought after all in comparison with acting and doing."[44] Does this mean, then, that thinking is not an activity, or—provided that it is a peculiar form of action or could become one—that it would no longer be conceptual thought? It is difficult for us to understand how a form of thought attempting to think ahead into the truth of being (the "meaning" of "being") can at the same time surpass all theoretical contemplation and knowledge as well as all practical activity and production. "Thus, thinking is an activity.

---

41   Marx, "Thesen über Feuerbach," no. 1. EN: "Concerning Feuerbach," 423.
42   Marx, "Thesen über Feuerbach," no. 8. EN: "Concerning Feuerbach," 423.
43   Heidegger, *Aus der Erfahrung des Denkens*, 9. EN: GA, vol. 13, 77; "The Thinker as Poet," 5.
44   Heidegger, *Über den Humanismus*, 6. EN: GA, vol. 9, 314; "Letter on 'Humanism,'" 240.

But an activity that at the same time surpasses all praxis."[45] Why can't we set off on our way, inspired by a form of thought which is neither theoretical nor practical—i.e., more fundamental than this distinction—, whilst considering "that there is a thought more strict than conceptual thought"?[46] Is it truly impossible for us to liberate ourselves from the instrumental use of "thought," from the institutional and cultural business of thinking, and unite with its most considerable aspects?

When Sartre defines his Existentialism as humanism and brings it into association with Marxism (within the empire of "-isms"), he simultaneously views existence as entirely rootless and restrictive, the humanity of humankind too anthropologically, and fails to grasp the essence of Marxism. He is unable to enter a dialogue with Marxism, as he remains caught in sundry pros and contras, merely occupies himself with particularities, and is not in a position to perceive the greater whole. For this reason, Heidegger writes to Jean Beaufret, using specific terms from Sartre: "Thought is not only *l'engagement dans l'action* for and through beings [*Seiendes*] in the sense of the reality of the current situation. Thought is *l'engagement* through and for the truth of being. Its history is never past, it always stands shortly before. The history of being bears and determines every *condition et situation humaine*."[47]

Therefore, the matter at hand is destiny, our destiny, the history of being, world history (and not universal history). Who among us does not know that for Marx the world—the true, real, effected—world is to be viewed as "the sensual world, as the total living sensual *activity* of the individuals constituting"

45   Heidegger, *Über den Humanismus*, 45. EN: GA, vol. 9, 361; "Letter on 'Humanism,'" 274.

46   Heidegger, *Über den Humanismus*, 41. EN: GA, vol. 9, 357; "Letter on 'Humanism,'" 271.

47   Heidegger, *Über den Humanismus*, 56 [EN: a typo for 5–6]. EN: GA, vol. 9, 314; "Letter on 'Humanism,'" 240.

68   it?[48] The extrasensory belongs to the superstructure and is determined by religious and ideological alienation, it is an "ideal supplement" to the real world. Heidegger is also aware of this and can even state: "The extrasensory becomes an empty by-product of the sensual."[49] This "degradation ends in meaningless-ness."[50] The extrasensory world is, however, the metaphysical world; "If, then, the essence of nihilism is borne in a history in which the appearance of beings as such as a whole [*das Seiende als solches im Ganzen*] in the truth of being does not occur, and accordingly, being itself and its truth are nothing, it follows that metaphysics—as the history of the truth of all things that have being as such—is nihilism in its very essence. If metaphysics is the complete historical foundation of Occidental and European world history, then this history is nihilistic in an entirely different sense."[51] Nihilism as the empire of the mastery of the will to will—already Nietzsche saw that: the will "would rather will *the nothing* than not will"—, and as the epoch of technology (itself understood as "metaphysics in completion"), perhaps forms the core of Marxism, its driving truth. A metaphysics both com-plete and "overcome," which converts to technology and thus—changed and transformed—"returns" and remains in dominance, becomes one with nihilism. "Nihilism is the world-historical movement of the peoples of the earth drawn into modernity's sphere of power."[52] But it is not "only" that. "Nihilism means: with everything in every respect, the nothing is going on. 'Everything' means beings as a whole. However, a being exists, when it is experienced as a being in each of its aspects. Nihilism means that

48   Marx, *Die deutsche Ideologie*, 42. EN: *The German Ideology*, 59.
49   Heidegger, "Nietzsches Wort 'Gott ist tot,'" 193. EN: GA, vol. 5, 209; "The Word of Nietzsche: 'God is Dead,'" 157.
50   Heidegger, "Nietzsches Wort 'Gott ist tot,'" 193. EN: GA, vol. 5, 209; "The Word of Nietzsche: 'God is Dead,'" 157.
51   Heidegger, "Nietzsches Wort 'Gott ist tot,'" 244. EN: GA, vol. 5, 264; "The Word of Nietzsche: 'God is Dead,'" 197.
52   Heidegger, "Nietzsches Wort 'Gott ist tot,'" 201–202. EN: GA, vol. 5, 218; "The Word of Nietzsche: 'God is Dead,'" 163–64.

it is nothing in relation to the whole of beings as such. But beings
[*Seiendes*] are what they are and how they are due to being [*Sein*].
Assuming that all that 'is' is so due to being, then the essence of
nihilism consists in its being nothing in relation to being itself.
Being *itself* is being in its *truth*, and this truth belongs to being."[53]
Nevertheless, the experience of being itself—its meaning and
truth—has never occurred. Not with the pre-Platonic thinkers,
either. "The history of being begins—and indeed necessarily—
with the forgetfulness of being."[54] This is itself an event—but not
the event [*Ereignis*]—and it is not merely an omission; the destiny
of being, within the oblivion of being, should not be attributed
to a diminished faculty of human thinking. Being itself conceals
itself, detaches itself, and remains absent. However, this absence
always takes place by virtue of "a presence and is determined by
this presence,"[55] and presence itself pervades this transcendence,
transcending that which disappears. The nothing that negates,
the nothing that belongs to being—and not merely the nothing
that is void and destructive and annihilated—, "much rather
affirms [. . .] itself as an exceptional presence, and veils [. . .]
itself as such a presence."[56] The nothing—which negates, which
even becomes present as "the other" to all things which have
being, as non-being, essences "as being"[57]—is more fundamental
than this other negation as well as every "not." As absence, the
nothing disrupts presence [*Anwesen*], negates it, but does not
annihilate it. History is and remains the history of the oblivion

53   Heidegger, "Nietzsches Wort 'Gott ist tot,'" 245. EN: GA, vol. 5, 265–66; "The
     Word of Nietzsche: 'God is Dead,'" 198

54   Heidegger, "Nietzsches Wort 'Gott ist tot,'" 243. EN: GA, vol. 5, 263; "The
     Word of Nietzsche: 'God is Dead,'" 196.

55   Martin Heidegger, *Zur Seinsfrage* (Frankfurt am Main: Vittorio Klostermann,
     1956), 32. EN: "Zur Seinsfrage," GA, vol. 9, 413; "On the Question of Being," in
     Pathmarks, 312.

56   Heidegger, *Zur Seinsfrage*, 23. EN: GA, vol. 9, 402–3; "On the Question of
     Being," 304

57   Martin Heidegger, *Was ist Metaphysik?*, 5th ed. (Frankfurt am Main: Vittorio
     Klostermann, 1949), 41. EN: "Nachwort zu 'Was ist Metaphysik?'," in GA, vol. 9,
     306; "Postscript to 'What is Metaphysics?'," in *Pathmarks*, 233.

of being, and leads into the beginning planetary consummation of nihilism, with which the final phase of nihilism only and first begins. Being itself, its truth and its meaning, be-ing [*Seyn*], has always remained forgotten. The being of beings remains absent and even undergoes annihilation in the oblivion of being, which is only now beginning to reach completion. But now, the truth of beings also perishes and a world characterized by metaphysics collapses. "The decline of the truth of beings implies: the revealability of beings, and *only* beings lose the previous singularity of their authoritative claim."[58] Nihilism negates and annihilates being that has always been forgotten, being that has been "crossed out," but simultaneously "the essence of the nothing in its former kinship with 'being,'"[59] and consummates the decline of the truth of beings. The nihilism that pervades being on the wrong track is not, however, "untrue," it is not by any means a going astray. "The essence of nihilism is neither curable nor incurable [*heilbar noch unheilbar*], it is the heal-less [*das Heil-lose*], but as such a unique reference to the salutary [*ins Heile*]."[60] In any event nihilism—as the "normal state" of humanity—cannot be overcome exclusively through "reactive attempts against [it] that strive for the re-establishment of previous circumstances instead of coming to terms with its essence."[61] Overcoming nihilism can only become an option subsequent to its realization, its consummation, and the beginning of its final phase. This end and the new beginning are not yet perceptible. Perhaps we are as of yet far removed from an experience of the truth of nihilism. In this context, the truth of nihilism means: that which corresponds to the essence of the oblivion of being, the collapse of the world, the

58 Heidegger, "Überwindung der Metaphysik," 72. EN: GA, vol. 7, 70; "Overcoming Metaphysics," 85.

59 Heidegger, *Zur Seinsfrage*, 29. EN: GA, vol. 9, 410; "On the Question of Being," 309–310.

60 Heidegger, *Zur Seinsfrage*, 9. EN: GA, vol. 9, 388; "On the Question of Being," 293.

61 Heidegger, *Zur Seinsfrage*, 13. EN: GA, vol. 9, 392; "On the Question of Being," 296.

essence of technology and labor, absolute productivity, the will to will. In nihilism, it is not only the being of beings which remains absent, being "crossed out" (being as being, be-ing [*Seyn*], rests from its essential origins within the oblivion of being); it is not only void (in the sense of the detached nothing as well as the nothing that negates and yet belongs to being) in relation to this being; the nothing itself remains absent. Man as the "shepherd of being" and "governor of the nothing" experiences neither being nor the nothing, and this concealment conceals itself and cannot be reduced to the activities or omissions of humanity. The consummation and overcoming of nihilism could—much likelier than "being"—allow the nothing (which neither possesses meaning nor is meaningless) to manifest itself. Heidegger poses the question in the following way and attempts to respond to it: "Does the nothing disappear with the consummation, or at least, the overcoming of nihilism? Presumably, this overcoming will only then occur when the essence of the nothing in its kinship with 'being' can arrive and find refuge with us mortals, instead of the appearance of the nihilative nothing."[62] Dare we comprehend the playroom of nihilism as a space reserved for the nothing? May one dare to perceive the play of time itself as an open horizon within which the *nothing-being* of *world-being* [*Nichts-Sein des Welt-Seins*], which is neither meaningful nor meaningless, neither extrasensory nor sensual, will act as a *game* and unfold its *errant truth* in a planetary manner?

Heidegger asks: "To where do being and the nothing belong, between which the *play* [emphasized by me] of nihilism unfolds its essence?"[63] We shall dare to make reference to the world game of nothing-being, within the dimension where nihilism is overcome, and to the play of time of true errance as the possibility of an entirely new clearing of being, as the possibility of an open

62   Heidegger, *Zur Seinsfrage*, 29. FN: GA, vol 9, 410; "On the Question of Being," 309–10.
63   Heidegger, *Zur Seinsfrage*, 32. EN: GA, vol 9, 412; "On the Question of Being," 312.

world that will survive the consummation of nihilism and persist beyond it. The "game of mirrors of the world" which Heidegger refers to, from which things become, become present, occur, and manifest themselves,[64] could be—as a world game—this open temporal space where being (even in its truth) and the nothing have negated themselves. The experience of absence is perhaps the fundamental experience of tomorrow. This new possibility of a clearing of nothing-being, this possibility of a new and open *world-being*, this possibility of an understanding of being within the game is perhaps already a necessity, a possibility corresponding to world exigency. After nihilism has not only voided the being of beings [*Sein des Seienden*], but even beings [*Seiendes*] as well, if we are prepared to realize that no previous unfolding of truth and no revelation of the world as a whole was *true*, and also could not remain as it was, we shall also be ready to put everything at stake against the horizon of the world play of world time.

V

Our topic was and is: Marx and Heidegger. And we arrived at nihilism. Why did we arrive at nihilism? How did nihilism come to us? On the way to Marx, on the way to Heidegger? Or is nihilism itself the main road? It could appear that we had forgotten Marx somewhat in favor of Heidegger. But is that actually the case? Do not externalization and divestiture, objectification, and alienation, just as much as the oblivion of being, homelessness, and unconditional objectification of beings, belong to the essence of nihilism? At the same time, does not the practical attempt to resolve alienation belong to the essence of nihilism, whilst even consummating the latter and allowing it to first achieve dominance in its unconditional "truth"? Did even Marx know about this when he realized that an opposite remains bound to

---

64   Martin Heidegger, "Das Ding," in *Vorträge und Aufsätze*, 180. EN: GA, vol. 7, 183; "The Thing," in *Poetry, Language, Thought*, 182.

its antithesis? "The supersession of self-estrangement takes the
same course as self-estrangement."[65] This thought occurred to
Marx and stands before us now.

Neither Heidegger nor this work attempt in the least to degrade
Marxian thought. Is it not Heidegger who gives the history
of being—of truth—expression while commemorating the
oblivion of being? Being that gives of itself and denies of itself,
that culminates in destiny, does it not achieve manifestation in
sundry ways? "With its inversions through Marx and Nietzsche,
absolute metaphysics belongs to the history of the truth of
being. Whatever originates from it, cannot be affected or much
less eliminated by refutation. [. . .] All refutation in the sphere of
essential thought is foolish. The dispute between the thinkers
is the " 'lovers' quarrel' of the matter itself," is a quotation from
the "Letter on 'Humanism.'"[66] Marx inverts metaphysics with
metaphysical means, and thus remains within the domain of the
dichotomy: sensual-extrasensory; Marx does not transcend the
concept of subjectivity; he even over-generalizes it. At the same
time, however, he opens another space. It would not be correct
to maintain that—in the case of Marx and Heidegger—different
interpretative approaches to the same circumstances can be dis-
cerned. All the same, the circumstances remain obscure.

The dialogue between Heidegger and Marx, provided that such
a dialogue is at issue here, requires an appropriate space and
the correct time. If we intend to subjugate this dialogue to the
polemic "Heidegger and Marxism," it will probably not be possible to find an exit, let alone a way that leads to the asking of the
question. It is a fact that Heidegger says nothing about class conflict, about the proletariat, about capitalistic exploitation. He says
nothing in favor of them and nothing against them. Just as he says

---

65   Marx, "Nationalökonomie und Philosophie," 232. EN: "Economic and
     Philosophical Manuscripts," 345.
66   Heidegger, *Über den Humanismus*, 23–24. EN: GA, vol. 9, 336; "Letter on
     'Humanism,'" 256.

74   nothing about sexuality and eroticism. Can we not read in *The Essence of Reasons* in regard to the neutrality of Dasein in relation to "sexuality [*Geschlechtlichkeit*]": "All essential propositions of an ontological analytic of Dasein in humans regard this being [*Seiendes*] in advance in its neutrality"?[67] Heidegger also attempts to think through so-called societal or political neutrality. He knows that the "private existence" bent on staying away from the "dictatorship of the public arena" does not at all coincide with the "free human being." Setting aside and withdrawal to the private sphere remain rigidly dependent—in a peculiar fashion—upon that which has been rejected, and receive their nourishment with reluctance, in servitude, from dominance. The event of encounter within affiliation does not have its roots in the ego or in "the they [*das Man*]"—and not in the connection between being and human. "If humans wish to come into the nearness of being once again, they must first learn to exist in namelessness. Similarly, they must learn to recognize both the seduction of the public as well as the powerlessness of the private. Before a word is uttered, the human must first be spoken to by being again, and perceive the risk that this aspiration may entail having little or seldom anything to say."[68] These "once again" and "again" do not imply a mere return to the past, as little as an unhistorical recurrence, a repetition. Marx also characterizes the process of historical movement "as reintegration or the return of the human as such."[69] A return to the origins, an origin that was never factual in the sense of an original situation played out in some past time, means: a leap into the future. Often, Heidegger has no term for this leap. In his way of proceeding, Heidegger neither evades the voice of stillness nor keeping silent.

67   Martin Heidegger, *Vom Wesen des Grundes*, 4th ed. (1929; Frankfurt am Main: Vittorio Klostermann, 1955), 38. EN: "Vom Wesen des Grundes," in GA, vol. 9, 158; "On the Essence of Ground," in *Pathmarks*, 122.

68   Heidegger, *Über den Humanismus*, 9–10. EN: GA, vol. 9, 319; "Letter on 'Humanism,'" 243.

69   Marx, "Nationalökonomie und Philosophie," 235. EN: "Economic and Philosophical Manuscripts," 347.

Do superiority or shortcomings conceal themselves in this
thought? Have we said anything when we answer: neither-
nor? Do we progress anywhere when we attempt to transcend
the question: either-or through our very questioning? Has the
question been resolved with the claim: Heidegger opens a
horizon within which class conflicts and sexual encounters, past
occurrences and preparations for the future are acted out and
squandered? Is *justice* a concern for Heidegger? Which justice?
One that we do not yet understand: "In order to prepare an
understanding of justice [. . .] we must neutralize all conceptions
about justice that originate from Christian, humanist, Enlight-
enment, bourgeois, and socialist morality," is a passage from *Off
the Beaten Track*.[70] Does this imply then that Heidegger wants to
surpass Marxism and socialism, or does he even fail to penetrate
their problematical essence? Does he stand on this side or that
side of the issue and which way does he go? All of this appears
to be so completely inaccessible through a given methodology.
The method—but which one?—is it a way, a *methodos*? Is dialectic
the most exceptional of methods? Marx did not mention dialectic
very much at all. He did, however, think dialectically, is the
response one receives to this. But what does *dialectically* mean?
Does anybody exist who did not think "dialectically" and was also
considered a thinker? By interpreting Hegel, Heidegger necessarily encounters the question of dialectic. He states the issue
as follows: "Likewise, the problem may be left undecided whether
dialectic is merely a mode of perception of whether it belongs to
objective reality itself as something real. The problem remains
a pseudo-problem as long as it has not been determined what
the reality of the real consists of, to the extent this reality lies
within the being of consciousness, and what the circumstances
of such being are. The elaborations about dialectic are akin to
the process of explaining a surging spring on the basis of the
stagnant water of a sewer. Perhaps the way to the spring is still

70   Heidegger, "Nietzsches Wort 'Gott ist tot,'" 157. EN: GA, vol. 5, 246–47; "The
      Word of Nietzsche: 'God is Dead,'" 184.

far off."[71] Therefore, dialectic cannot be adequately interpreted in terms of the motion: thesis, antithesis, synthesis (or: position, negation, negation of negation), nor as infinite negativity. It is neither subjective nor objective, neither logical nor ontic. What is it? That which remains to be comprehended from *logos* and the "dialogue," from the correspondence to the beckoning of being, from language (*logos*) and contradiction, from dispute and antagonism, from "being" and becoming, from subjectivity and material substantiality—and their dynamic, from implication and confusion. Dialectic, "however, as a dialectic of the history of being, transmuted into historical dialectical materialism, determines the history of humanity today in manifold ways. The world historical conflict [*Auseinandersetzung*] of our era has much older origins than the political and economic power struggles in the foreground would like us to think."[72] Whatever is in the foreground does not for this reason lose its relation to its underlying foundation. Its underlying foundation [*Grund*] and abyss [*Abgrund*]. Then: "Precisely this is what appears as that which must now be thought, namely: being 'is' the abyss [*Ab-grund*], and accordingly, being and ground [*Grund*] are the same. To the extent being 'is' grounded, and only to this extent, it is without any ground."[73] Would then being cease to have any foundation in order to be thrust into the game as being? The vicissitudes of destiny and of the evasion of being—could they be taken up by the world game itself and dissolved within the play of time? The "why" question—could it be converted into the questioning answer "without why," neither groundless nor grounding, neither in the tone of a tragedy nor in the style of a comedy—without

71 Martin Heidegger, "Hegels Begriff der Erfahrung," in *Holzwege*, 168. EN: "Hegels Begriff der Erfahrung," in GA, vol. 5, 183; "Hegel's Concept of Experience," in *Off the Beaten Track*, 137.

72 Martin Heidegger, *Der Satz vom Grund* (Pfullingen: G. Neske, 1957), 149–50. EN: GA, vol. 10, *Der Satz vom Grund*, 131; *The Principle of Reason*, trans. Reginald Lilly (Bloomington: Indiana University Press, 1996), 88.

73 Heidegger, *Der Satz vom Grund*, 185. EN: GA, vol. 10, 166; *The Principle of Reason*, 111.

having to explain through beings [*Seiendes*] the openness of the
open horizon?

Do the unleashed production and reduction of the world
(*genitivus subjectivus* and *objectivus*) proceed "dialectically"? Or
rather, does dialectic provide us with serious ontic explanations
or a playful ontological understanding of the world? And how
does dialectic function within the whole ontic-ontological
relation? Every understanding of dialectic—even when it is
endeavored knowingly—remains problematical. Heidegger states
rightly so: "prepared by Kant, thought has been brought in certain
aspects to the highest dimension of its possibilities through
the efforts of the thinkers *Fichte*, *Schelling*, and *Hegel*. Thinking
becomes knowingly dialectical. Moving within this dialectical
sphere, and indeed, stirred even more by its unfathomed depths,
are the poetic minds of *Hölderlin* and *Novalis*. The theoretical and
speculative, thoroughly executed unfolding of dialectic within
the completeness of its purview is carried out in Hegel's work
with the title *The Science of Logic*."[74] Nevertheless, the question still remains open: *which* dialectic—i.e., which language
and which thought, which logic and which reality and in which
relation and in which differentiated unity—comes to dominance?
"Admittedly, as soon as dialectic is the topic, someone notes that
there is a dialectical materialism. One takes it for a worldview,
treats it as an ideology. But with this ascertainment we neglect
deeper reflection on the matter, instead of recognizing that:
today, dialectic is a reality, perhaps even the reality of the world.
Hegel's dialectic is one of the thoughts—brought forth from
afar—'that direct the world', with equal significance there where
dialectical materialism is believed, as well as there where—only

---

74   Martin Heidegger, "Grundsätze des Denkens," *Jahrbuch für Psychologie und
Psychotherapie* 6, no. 1/3 (1958): 34. EN: GA, vol. 79, *Bremer und Freiburger
Vorträge: 1. Einblick in das was ist 2. Grundsätze des Denkens, 82; Bremen and
Freiburg Lectures: Insight Into That Which Is and Basic Principles of Thinking*,
trans. Andrew J. Mitchell (Bloomington, IN: Indiana University Press, 2012),
78.

in a somewhat altered style of the same thought—it has been refuted. Behind this, as one says: ideological conflict, the war for world dominance rages on. Behind this war, however, a dispute persists through which Occidental thought remains at odds with itself. Its final triumph, to which it is beginning to rise, consists in the fact that this thought has forced nature to surrender nuclear energy."[75] The whole and holistic—and simultaneously fragmentary—"dialectical" play of world production and reduction cannot be comprehended by either the dynamic of thought and reflection, or by concrete motion and activity, although Heidegger is able to write: "in an earlier work published from his posthumous writings ['*National Economy and Philosophy*'],[76] *Karl Marx* states that 'the *entire so-called history of the world* is nothing other than the generation of humanity through human labor, as the becoming nature of humans.' [. . .] Many will reject this interpretation of world history and the concept of the essence of the human underlying it. But nobody can deny that—as the work of the self-production of humanity—technology, industry, and economics decisively determine the actuality of the actual today. Already this determination removes us from the dimension of thought within which the quoted passage from *Marx* about world history as 'the labor of the self-production of the human' is moving. For in this context the word 'labor' does not designate mere activity and performance. The word speaks in the sense of *Hegel's* concept of labor, which is conceived as the fundamental factor of the dialectical process that allows the becoming of the real to unfold and complete its actuality. The fact that *Marx*, in contrast to *Hegel*, does not see the essence of reality in an absolute spirit in the process of comprehending itself, but rather in humanity producing itself and its means of sustenance, does indeed place *Marx* in stark contrast to *Hegel*, but this contrast allows *Marx* to remain within Hegelian metaphysics; for the life

---

75   Heidegger, "Grundsätze des Denkens," 37. EN: GA, vol. 79, 88; *Bremen and Freiburg Lectures*, 84.

76   EN: The additions in brackets are from Axelos.

and workings of reality is everywhere the process of labor as
dialectic and, i.e., as thought to the extent the actual productive
element of each production remains thought, whether the
thought in question is carried out in a speculative and metaphysical mode, in a scientific and technical mode, or a mixture and
banalization of both. Every production is intrinsically re-flection,
is thought."[77] With separation or connection, with the relation to
being or the affiliation between logos, theory, thought, reflection,
consciousness and praxis, *techne*, activity, reality, action, being—
i.e., with the question about "dialectic," both Marx and Heidegger
have their difficulties. In his polemical writing against Proudhon's
Philosophy of Poverty,[78] composed in French, and although Marx
is aware that labor represents a certain unity of "real" and
"ideal" forces—within a matrix of interrelations—, he speaks of
"mouvement réel de la production" and makes ironic reference to
dialectical abstractions, products and theoretical concepts and
pretenses of objective—practical and material—movement.[79]
And in his "Introduction" to the *Contribution to the Critique of
Political Economy* he emphasizes the fact that the origins of
modern economics are first realized in the most modern forms
of existence [*Dasein*] within bourgeois and capitalist economics:
"Here, therefore, the abstraction of the category 'labor', 'labor as
such', labor sans phrase, the initial point of modern economics,
becomes objectively true for the first time."[80] At the same time,
Marx knows that efficacious labor cannot be separated from

77   Heidegger, "Grundsätze des Denkens," 40–41. EN: GA, vol. 79, 94–95; *Bremen
     and Freiburg Lectures*, 90–91.
78   EN: Pierre-Joseph Proudhon, *Système des contradictions économiques ou
     Philosophie de la misère* (Paris: Guillaumin & Cie, 1846); *System of Economic
     Contradictions: Or, The Philosophy of Misery*, trans. Benjamin R. Tucker
     (Boston: Benj. R. Tucker, 1888).
79   Karl Marx, *Misère de la philosophie* (1847; Paris: Costes, 1950), 51, 121–22,
     125–27, 148–49. EN: *The Poverty of Philosophy*, London: Martin Laurence
     Limited, 1937. I have been unable to find a copy of the specific edition Axelos
     used. My best guess for the relevant passages is 43, 91–92, 94–96, 112–13.
80   Karl Marx, *Zur Kritik der politischen Ökonomie* (Berlin: Dietz, 1951), 261. EN:
     Grundrisse: Foundations of the Critique of Political Economy (Rough Draft),

80 thought all that simply: "What qualifies the worst architect from the best bee from the onset is the fact that he has built the cell in his head before it is done in wax. At the end of the process of labor, a result is produced that already existed from the beginning in the *mind of the worker*, and thus, was already present *ideally*. Not that the worker has *effected* merely a change in form of the natural object; the worker *realizes* at the same time *his purpose* through the natural object, a purpose he knows and which determines the method of his action in the way of a law, and to which he must subordinate his will. And this subordination is not an individual act. Besides the exertions of organs that must function, the *purposeful will*—which expresses itself as *attention*— must be commanded for the entire duration of the labor, and all the more, the less this labor fully occupies the worker by virtue of its content and the method of its execution, and thus, the less the worker enjoys this labor as the *play* [emphasized by me] of his own bodily and mental powers."[81] Must, then, the way be found in the direction of a *techno-logos* uniting reality "and" thought, work and play? "Technology reveals the active behavior of humans to nature, the immediate process of production of their life, and hence, of the circumstances of his social life and the intellectual conceptions arising from them, as well."[82] A neither idealistic nor materialistic technology, which is capable of manifesting itself even meta-dialectically, does not exist.

In the final hour of his lectures about "The Principle of Reason" (1955/1956), which also contained the passage concerning "historical and dialectical materialism"—the latter should not be regarded outside the context of other statements made about materialism in the lecture "The Principle of Reason," namely, that the specter of materialism does not move all that one-sidedly, as: "It does not sweep in from the west with any less force than

trans. Martin Nicolaus (London: Penguin in association with New Left Review, 1973), 104–5.

81   Marx, *Das Kapital*, vol. 1, 186. EN: *Capital*, vol. 1, 284.
82   Marx, *Das Kapital*, vol. 1, 389. EN: *Capital*, vol. 1, 493, n. 4.

from the east"[83]—, Heidegger concludes this lecture with a query about the game. He gives the game, in which being rests as being without any underlying foundation, the chance to be expressed and to become a matter of consideration. Not only in the play-of-time space [*Zeit-Spiel-Raum*] where beings appear and are produced. As a game without a "why," as the highest and the deepest, as the destiny of being, as the one-all—perhaps this is how we must conceive of the meaning of being without imagining the game "as something that exists." "Can the essence of the game be measured properly on the basis of being as ground, or must we conceive of being and ground, being as a-byss on the basis of the essence of the game, indeed, a game into which we mortals have been brought, the mortals that we are as those who dwell near death, which is the most extreme potential of Dasein, and thus, capable of attaining the highest in respect to the clearing of being and its truth?"[84]

As a result, thought is put at risk, inextricably tied to the game. Our thought is not yet able to conceive of the game. Will planetary thought have the ability to think truly-errantly the world (the worlding world [*die weltende Welt*]—the clearing and darkness of being) as a game, that worlds and times? The worldly time of world-being—will it reveal itself in the game? The game as the one, the open whole, the one-all, the sole and unifying—could it be played, too, but not only with beings? Because it is playing?

We humans, will we succeed in entering the game in a harmonious and planetary manner, and how can we—put at risk—live playfully and perish?

Against this horizon, the difference between the thought of Marx and Heidegger disappears. However, as a difference that has been taken on and resolved, it remains at risk nonetheless.

---

83   Heidegger, *Der Satz vom Grund*, 199. EN: GA, vol. 10, 179; *The Principle of Reason*, 122.

84   EN: Heidegger, *Der Satz vom Grund*, 186–87; GA, vol. 10, 167; *The Principle of Reason*, 112.

Perhaps the world game also revealed itself to Marx as a game: after the resolution of the alienation of labor, and even after communism has been overcome, couldn't the history of humanity—and not only this—manifest itself as a game, a game in which the inviolable essence of alienation would also play a role? Whilst neither remaining captive in the realm of necessity nor comprehended as freedom, whilst neither conceived of as a unifying dialectic nor as something arbitrary, could this game hold together everything that exits in a diffuse cohesion? Without the game itself, which is taking place, being something to be found in the midst of all beings [*Seiendes*]?

VI

The human has not yet found a position in the game of the world. Must one become a "citizen of the world" beforehand? Must one join the world game as a world-citizen? Marx demands the existence of the human within the context of world history: "The existence of individuals in world history, i.e., the existence of individuals which is immediately connected with history."[85] Here, world history does not yet mean the destiny of world-being; the citizen of the world always exists without a definite form and they will always remain that way. World history and the road to world citizenship perhaps lead—pointing beyond themselves—to a world of openness, to a play-of-time space with different possibilities. In the world game of world-being and the play of time, could a new form of humanity be at home there, or find its home there? Each homeland would not be patriotic, nationalistic, international, or cosmopolitan; it would exist in the sense of the history of being and the world, in the sense of destiny, if it should "be" anything at all. Indeed, the wandering star within the play-of-time space has become exactly what it is: a planet. Cannot humans, too, first inhabit what their place is—within the play-of-time space? With nostalgia and the yearning for faraway

85   Marx, *Die deutsche Ideologie*, 33. EN: *The German Ideology*, 47.

places? Provided that adventure and the return home can still be distinguished from one another.

In order to find a way through homelessness and reach the world homeland, must the human of the modern epoch of subjectivity (which merely becomes generalized through collectivity) and the objectification caused by it be overcome beforehand? Marx still belongs to this epoch: he views the human as the *animal rationale* of metaphysics, the rational living being that is driven by its (biological) drives and strives to satisfy them with material means, whereby technology and thought are utilized to achieve this purpose. Correspondingly, he himself characterizes his thought as naturalism-humanism-socialism in completion. In departure from the animalistic—although human and social—nature of humanity, he strives for the humanization of beings through the human praxis of objectified subjectivity, which should lead to a unified communistic society of man. The Cartesian ego of the *cogito*, which means the *res cogitans,* the transcendental subject of Kant and the transcendental substantiality, Hegel's absolute subject as absolute knowledge, as the will of the spirit, becomes in Marxian thought. "the subjectivity of essential objective powers whose actions are therefore also necessarily *objective*,"[86] which he wishes to raise to the level of the all-encompassing and all-justifying societal (the social-socialistic level). In the "Letter on 'Humanism,'" Heidegger, "on the contrary," demands that the naturalistic, psychological, sociological, and humanistic conception of humans be overcome through a way of thinking that dares to ask in which respect the essence of humans belongs to the truth of being. His "objection" to humanism is that it does not value the humanity of humankind, its essence, its nature, highly enough, and does not place it in the clearing of being. He requires another category of essence in order to experience "how the human in their proper essence

86   Marx, "Nationalökonomie und Philosophie," 273. EN: "Economic and
      Philosophical Manuscripts," 389.

becomes present to being"[87], in the ecstatic interiority within the truth of being, in ex-istence, in Da-sein. "Through this category of human essence"—Heidegger stresses—"humanistic interpretations of the human as animal rationale, as 'person', as an intellectual-spiritual-corporeal being are not declared to be false and not rejected. Much rather, the sole thought is that the highest humanistic determinations of human do not yet achieve the appropriate dignity of the human."[88] The subject of humanism, the subject of homelessness, of alienation, of subjectivism and objectivism, never encounters themselves; their essence remains alien to them. The human who conceives of themselves as the ruler of being, as the subject of beings [*Seiendes*] (of the object), remains "standing" in the midst of beings; they fall victim to the thoughtlessness that condemns one with such a self-conception (i.e. misconception of self) to remain hanging in the midst of beings. Humanism makes humans incapable of contemplating that and how they are "thrown" into the world by being with "their" being-in-the-world [*In-der-Welt-sein*]; in this way, however, the human cannot be brought to ex-istence where they can tend to the truth of being and respond to the nothing as its governor [*Platzhalter*].

This objection is not formal and not critical; it is directed at the affiliation between the essence of being and human. It strives for the liberation of human "and" of the truth of being. "One consistently thinks of *homo animalis*, even when *anima* has been posited as animus sive mens, and later, the latter as subject, as person, and as mind. Such positing is characteristic for metaphysics. But in the process, the respect shown here for the essence of man is too low and its origins are not taken into consideration, for the origins of the essence of historical humanity always remain the future of that essence. Metaphysics conceives of the human

---

87   Heidegger, *Über den Humanismus*, 19. EN: GA, vol. 9, 330; "Letter on 'Humanism,'" 251.

88   Heidegger, *Über den Humanismus*, 19. EN: GA, vol. 9, 330; "Letter on 'Humanism,'" 251.

on the basis of *animalitas* and does not progress to the thought
of its *humanitas*."[89] Do these statements made by Heidegger
concern those things that were thought by Marx? Accordingly,
we hear from Marx: "The human is most directly a *natural being*.
As a natural being and living natural beings they are—partly
equipped with *natural powers*, with *vital forces*—an *active* natural
being; these powers exist within them as assets and skills, as
drives;[90] in part, as a natural, corporeal, sensual, and material
being, man is a *suffering*, conditioned, and restricted being, just
like an animal or a plant, i.e., the *objects* of their drives exist outside of themselves, as *objects* independent of themselves. But
these objects are *objects* of their needs, they are *objects* indispensable and essential for the actualization and affirmation
of their essential powers."[91] This simultaneous positing of the
object as an object for a subject as well as the human subject as material, this natural, corporeal, biological, concretely
subjective-objective positing of the "essence" of the human,
is characteristic of Marxian humanism, as naturalism in completion. The acknowledgement of the natural, human, and social
human demanded by Marx—which we have just encountered in
the quotations from Heidegger—, who should be able to satisfy
all of their natural, human, and social needs through technology
unchained within the realm of communism, is argued on the
basis of an objective positing of a subject, a subject coerced
by their drives, in order to achieve the concrete socialization
of subjectivity (in collectivism). Does, then, Marx also belong
to the epoch of subjectivity, the metaphysics of humanism?

89   Heidegger, *Über den Humanismus*, 13. EN: GA, vol. 9, 323; "Letter on
     'Humanism,'" 246–47.
90   Freud and psychoanalysis also expound in a productive manner on the
     naturalistic and "natural," i.e., biological and psychological conception of
     instinct used by anthropology. At the same time, however, Freud releases
     the power of desire for life, love, destruction, and death, and sets the play
     of the unconscious, the preconscious, and the imaginary into incredible
     motion.
91   Marx, "Nationalökonomie und Philosophie," 274. EN: "Economic and
     Philosophical Manuscripts," 389–90.

He does indeed radicalize the dynamic whose foundations allow metaphysics (the extrasensory) to convert to physics (the sensual), although—or perhaps because—he writes the following (which is correct but also restrictive): "However, life requires above all food and drink, shelter, clothing, and other things [is this the elementary?]. The first historical act is, therefore, the generation of means to satisfy these needs, the production of material life itself, and this is indeed an historical act, the fundamental condition of all history, which must still be fulfilled daily and on an hourly basis today, just as it was centuries ago, in order just to keep humans alive."[92] And for Marx, it is production, technology, that should ensure life on a continually broader foundation. Correspondingly, the entirety of beings should be transformed through technical and productive labor in order to satisfy the needs of man.

The space of "world-being," seen in the context [*Rahmen*] of beings [*Seiendes*] and only beings, must be fulfilled in this technical-productive manner. But can the emptiness of being be filled by production and technology? Being [*Seiende*] appears increasingly unsatisfactory, deficient, and technology addresses this deficiency with technical solutions. At the same time, it produces new needs in order to be in a position to satisfy them. There is increasingly more shortage, production increases more and more, and more and more is consumed and used. The entire earth-ball is drawn into this spiral and it appears to be completely consistent that this process is spreading and intensifying. Production appears to be the foundation of beings as well as their transcendence, and it is in fact the case that modern technology is a disclosure, a challenge, a fabrication. By disclosing and establishing beings in this way, is the humanity of modern and already planetary technology responding to the beckoning of unconcealment, or does it contradict the latter? Perhaps humanity corresponds to it by virtue of contradiction,

92   Marx, *Die deutsche Ideologie*, 24. EN: *The German Ideology*, 39.

and perhaps humanity sublates its subjectivity as well as its
characteristic substantiality in this way. Perhaps humanity
achieves a new openness on this path—if it is indeed a path—,
which overcomes objective subjectivity and objective substantiality, and which dissolves and transcends the subject-object
dichotomy so omnipresent for modernity. Something new is
added to the flux of reality. However, that which unconceals
itself is always a clearing that unconceals and conceals itself;
that which reveals itself withdraws at the same time. In a certain
sense, Marx and Heidegger (but not the latter in the way of the
former)—one on the basis of the emptiness of beings, the other
on the basis of the oblivion of being—strive for the same thing:
to expand our awareness of technology, technology as a destiny
but not as an unavoidable course and "fate." Without pursuing it
blindly, or—which would be the same thing—damning it as the
work of the devil, we should in fact open ourselves explicitly to
the essence of technology. For and only "if we open ourselves
explicitly to the *essence* of technology, we find ourselves unex-
pectedly taken up by a liberating claim."[93]

Everything that exists increasingly appears under the aspect of
reduction, as a deficient fullness, as the growing diminishment
of originality. However, this reduction is not the mere result of a
given perspective. It corresponds to the productive will to power,
it accompanies the latter and is also determined by the latter. Is
the will to power, the will that wills itself, the unconditional will
to will, a production of the being of beings? Is the production of
technology a production grounded on "being"? The duel between
reduction and *production* has not yet unfolded its game to its full
extravagance. We still stand much too much within the realm of
metaphysics, and we stagger for this reason. Of a metaphysics
that tells us "that and how the being of beings 'is' [*Daß und wie*

93   Martin Heidegger, "Die Frage nach der Technik," in *Vorträge und Aufsätze*, 33.
EN: GA, vol. 7, 26; "The Question Concerning Technology," in *The Question
Concerning Technology and Other Essays*, trans. William Lovitt (New York:
Harper & Row, 1977), 26.

'es' das Sein des Seienden 'gibt'],"[94] of a metaphysics as the destiny of transition (of "transcendence"), of the transition of the being of beings. "It almost seems to be the case that the way metaphysics conceives of beings obliges it to be the unknowing limit that denies man the original relation to the being of human essence."[95] According to metaphysics, all "objectivity" as such is "subjectivity"; in this way, the foundation of beings erodes, and the *onto*logical logic prevails over *onto*logical metaphysics in an empty—and overfilled—space. Thus, we do not experience being as being, nor that the nothing which belongs to being. Overcoming nihilism means: overcoming and recuperating from metaphysics, the preparation of a new horizon. The nothing still remains to be experienced. "Being and nothing belong together, but not because both—considered in terms of the Hegelian concept of thought—coincide in their indefinition and immediacy, but rather because being itself is finite within essence and only reveals itself in the transcendence of Dasein projected into the nothing."[96] The nothing becomes present as being. It "is" not, but it is also not nothing [*nicht nichts*]. "Being 'is' just as little as the nothing 'is.' But *There is* both [*Aber* Es gibt *beides*]."[97] Both in one In being-nothing [*Beides in einem Im Nichts-Sein*]? Against a certain horizon, wouldn't being and the nothing be the same? "Unlike beings, being does not allow itself to be conceived of and produced as an object. The 'other' to beings is in itself non-being [*Nicht-Seiendes*]. We renounce thought all-too-prematurely when we declare in a deficient manner that the nothing is the merely insubstantial and equate it with the absence of essence. Instead

94   Heidegger, *Zur Seinsfrage*, 33. EN: GA, vol. 9, 413; "On the Question of Being," 312.

95   Martin Heidegger, *Was ist Metaphysik?*, 11. EN: "Einleitung zu 'Was ist Metaphysik?,'" in GA, vol. 9, 370; "Introduction to 'What is Metaphysics?,'," in *Pathmarks*, 281.

96   Martin Heidegger, *Was ist Metaphysik?*, 36. EN: "Was ist Metaphysik?" in GA, vol. 9, 120; "What is Metaphysics?" in *Pathmarks*, 94–95.

97   Heidegger, *Zur Seinsfrage*, 38. EN: GA, vol. 9, 419; "On the Question of Being," 317.

of yielding to the haste of vapid astuteness and sacrificing the
perplexing ambiguity of the nothing, we must mobilize our
singular readiness to experience in the nothing the spaciousness
of what bestows beings with the guarantee to exist. This is being
itself."[98] Thus, the nothing becomes present as being. The nothing
is being itself. In *Off the Beaten Track*, it is said about the nothing
once again: "it is being itself."[99] Is this the highest game, the one,
the only, the all-unifying? The world game within worldly time?
The game plan overcome? We must not forget that being and time
cannot be separated. "In *Being and Time* 'being' is not something
different than 'time,' to the extent that 'time' is used as the first
name for the truth of being, as this truth is the becoming essence
of being, and hence, being itself."[100] Being, time, world, nothing,
game are—without being beings—the same reality: nothing-
being, world-time, world-being, the play of time, the destiny of
being and the world·

The production of technology unfolds as a provocation to being,
and thus, "also" to the nothing, it storms into world-being within
worldly time and the play of time. Marx reduces the world to the
productive aspects of technology; Heidegger demands that we
explicitly open ourselves up to the essence of technology·

It is necessary to respond to the provocation of being.
Provocation of being means: the challenge (of production)
brought forth by world-being itself to rise to meet this
productivity, this provocation, this challenge by technology,
and cope with it. In this duel, the provocative and productive,
playful cunning of world-being encounters that technology and

98   Heidegger, *Was ist Metaphysik?*, 41. EN: GA, vol. 9, 306; "Postscript to 'What is
      Metaphysics?,'" 233.

99   Martin Heidegger, "Die Zeit des Weltbildes," in *Holzwege*, 104. EN: GA, vol. 5,
      113; "The Age of the World Picture," in *The Question Concerning Technology
      and Other Essays*, 154.

100   Martin Heidegger, *Was ist Metaphysik?*, 16. EN: GA, vol. 9, 376; "Introduction
      to 'What is Metaphysics?,'" in *Pathmarks*, 285.

productivity challenging world-being and the nothing. Each of both forces is provocative and provoked.

It cannot yet be envisioned how this dispute will end; will everything decline in void the nothing and perish? Destroy nothing and perish? Will nihilism be overcome? If we succeed in seeking refuge in nihilism, in lifting everything that exists and was ever made into a higher ambivalence, in subjugating ourselves to the game as a game and playfully participating in it, will our endeavors be given the chance to dwell within the game holistically and disrupted, in a *harmonious and planetary* manner—within the play of time? Those statements made by Marx must first be elevated into this higher ambiguity and borne by it. Within this ambiguity, Marx and Heidegger encounter one another and distance themselves through the things that they say, and through that which is not thought and not pronounced. Above all, they have been set in motion and united in a "third" planetary phase beginning to unfold. The constellation of a dawning planetary age stands and moves beneath a higher motion of the stars.

The duel between man and world has not yet been unloosed by any means. Everything conceals itself, withdraws, appears as trivial. At the same time, other horizons are opening. Marx "like" Heidegger, each in a different language, gives expression to what that poet of this wandering star, who experienced his magnificent decline in madness , did indeed say—though in a romantic and utopian manner: *We are nothing; what we seek is everything.*

II: ON MARX AND HEIDEGGER

Theses on Marx: Concerning the Critique of Philosophy, of Political Economics, and of Politics

Today, the whole world knows, or imagines that it knows, what the situation is with Marx's undertaking. Marx wants to overcome vulgar, naturalistic, and mechanical materialism, which is obscured by its object and does not acknowledge the active subject that objectifies itself through the act of production. The *Theses on Feuerbach* provide the most ingenious critique of ahistorical and undialectical materialism. In his famous epilogue to the second edition of *Capital*, Marx writes: "My dialectical method is not only different from that of Hegel in respect to its foundations, it is also its direct opposite. In Hegel's view, the process of thought, which he even transforms into an autonomous subject with the name idea, is the demiurge of a reality that only constitutes its external appearance. In my view the reverse is the case—the ideal is nothing other than the material transformed and converted by the human mind . . . [Dialectic] is standing on its head with Hegel. One must upright it in order to discover the rational kernel within the mystical shell."[1] But what happens with thought according to this dialectic walking on its own feet? Hegel considered philosophy to be an *inverted world* and thought philosophizing implied *standing on one's head*; but in his opinion the world of naïve consciousness is in fact the wrong way round and inverted. Marx intends to right the inverted world, although he is aware that "the inverted world is the real world," as long as alienation has not been resolved. Might it be the case that the reality he demands is burdened by a realist conception of a perverted and inverted world?

Marx's conception strives to overcome philosophical thought, subsequent to its realization through concrete praxis, through the practical subject as active materialist. What happens to the *unity* of the world as a consequence? How should—during the process of overcoming—the truth of materialism and idealism be preserved at the same time? Will one of these two forces—both of which should be overcome—attain an absolute dominance?

It is useful not to view Marx as a "dead dog"—he himself demanded that one should not treat Hegel this way—,[2] but rather, one should question one's own theoretical, economic, and political thought and everything that refers to Marx; but neither in order to amass scholarly works on all kinds of Marxian, Marxist, and Marxological topics, as they solicit themselves to dispassionate research, nor in the sense of Marxian, Marxist, and Marxological pedantry whose dissemination has only just begun.

Since Lenin ("The Three Sources and Three Component Parts of Marxism"), Marxism is viewed as having borrowed from three sources—from *classical German philosophy*, from *English political economics*, and from *French utopian socialism*—with a sub-division into three constitutive parts: *philosophy* (of historical and

---

1   EN: Axelos does not provide a reference. The passage can be found in "Postface to the Second Edition," *Capital*, vol. 1, 102.

2   EN: Marx, "Postface to the Second Edition," 102.

dialectical materialism), *political economy* (based on the theory of
labor, value, and surplus value), and *politics* (of class conflict and
the dictatorship of the proletariat).[3] This is how the three fronts
of the ideological, economic, and political struggle were constituted. Perhaps it is time to allow negativity to become effective
within each of these three forces as well, and within the unified
center from which they grow.[4]

I

The main deficit of all previous versions of materialism (including
Marx's historical dialectic), is the fact that objects, reality, and
materials are merely comprehended in the form of the manufactured objects, the material realities, and the materials of labor;
they are in fact comprehended in this way, but they lack a *ground*
and *horizon*. Therefore, the other metaphysical perspective—a
contrast to the naïve or scholarly realism, which was developed
on the basis of idealist philosophy, and which of course neither
knows nor recognizes the so-called real world; the whole of
figures, forces, and weaknesses of the constituted, concretized,
and fixed world, the other side of the same and only world. Marx
desires sensual objects, but higher objects which are actually
distinct from objects of thought: but he does not comprehend
human activity itself as a *problematic activity*. For this reason, in
the *Contribution to the Critique of Political Economy* just as much
as in *The Poverty of Philosophy*, Marx considers material life to
be genuinely human, whereas thought and creative expression
are comprehended and posited merely in their contingent and
ideological form of appearance. Thus, he does not grasp the

---

3   EN: Vladimir Lenin, "The Three Sources and Three Component Parts of
    Marxism," in *Collected Works*, vol. 19 (Moscow: Progress Publishers, 1977),
    21–28; originally published in *Prosveshcheniye*, no. 3 (1913).
4   See Kostas Axelos, *Marx penseur de la technique: De l'aliénation de l'homme à
    la conquête du monde*, Collection "Arguments", 2nd ed. (1961; Paris: Éditions
    de Minuit, 1963). EN: Translated by Ronald Bruzina as *Alienation, Praxis and
    Techne in the Thought of Karl Marx* (Austin: University of Texas Press, 1976).

significance of a thought which places everything in question and keeps these questions open, a thought which dares to see that every great victory is the prelude to defeat.

II

The question as to whether *(constantly errant) truth* should be attributed to human thought, is neither a theoretical question nor a practical question. During this questioning, man must experience the truth, i.e., reality and unreality, power and failure of his thought and the world. The dispute about the reality or non-reality of praxis—which is isolated from *questioning thought*—, is a purely pragmatic question: it lacks a foundation that has been wrested from the abyss and confusion, it does not dare to shake the security of fixed positions.

III

The Marxist doctrine of changing concrete circumstances and education forgets that circumstances are not always coequal with human subjects, and that all educators mislead. Therefore, it must divide society into two parts—of which the one remains predominant. The asymmetry of changing concrete circumstances and of human change or self-alteration can only be comprehended and "truthfully" understood as a permanent revolution that places itself in question.

IV

Marx uses the fact of economic alienation, the bifurcation of the world into a world as (real) base and a world of (ideological) superstructure as a point of departure. His work consists in dissolving the ideological, idealist, and ideal world into its worldly foundations [*weltliche Grundlage*]. But the fact that these worldly foundations set themselves apart from themselves [*sich von sich selbst abhebt*], while an autonomous empire stretches across the

entire surface of the earth [*Erdoberfläche*], can only be attributed to the lack of coherence and grounding of these world
foundations. Therefore, this situation must be comprehended
and revolutionized in itself as well as in its inadequacy. For
example; after the earthly [*irdische*] family has been exposed as
the alleged secret of the holy family, the institutions of the former
must be opened out until they shatter.

V

Marx, dissatisfied with realist thought, desires praxis; but he conceives of praxis not as sensual activity, which leaves the *question
of meaning* undecided. In the midst of an epoch in which rampant
planetary technology wages a deadly war against the world,
would we not be in need of a *techno-logy* capable of thinking
everything encompassed by technology as well as technology
itself? For technology does not merely encompass those things
expressly belonging to the spheres of the universe and cosmology, of life and biology, of the dynamics of the psyche and
psychology, of society and sociology, of ideas and ideology; it
draws beings [*Seiendes*]—and what has been produced—into its
clockwork [*Räderwerk*].

VI

Marx dissolves worldly essence into human essence. But human
essence cannot rest upon itself as it could on its own foundation.
Within the play of becoming, it is the fragment of a dialogue
without which the human would neither strive to be *human* nor
would the world "be" *world* [*noch die Welt* Welt „*wäre*"].

Marx, who does not concern himself with this constellation of
human and world, is therefore forced:

> 1. to abstract from time that is always open, posit human
> essence using a specific practical and humanistic inter-
> pretation, and presume an objective human essence. He

thus loses sight of the human as the essence of distance [*So verliert er den Menschen als Wesen der Ferne aus den Augen*].

2. Accordingly, this essence can only be comprehended as an empirical totality, as a productive and produced generality which unites all essences in a single category with technical means.

VII

For this reason, Marx does not perceive that economic production itself is a product and that the empirical society he analyzes belongs to an *errant world*—in whose course everything violent and peaceful continues to play on in its disconcerting way; for critique belongs to the criticized.

VIII

All social life is essentially *in strife*. All puzzles which guide thought into openness cannot find their "truthful" solution either in human praxis or in the simplistic or scholarly and critical comprehension of this praxis.

IX

The apex of practical materialism, i.e., that materialism which conceives of thought as a derivative activity, is the abstract anticipation of the total society and the total human.

IX

The standpoint of Marxian materialism is the socialized society; but a new way of thinking—without a standpoint and without a one-dimensional (neither spiritualistic nor materialistic, neither idealist nor realist) orientation—must place its focus on the game

of the planetary world—for the becoming being of the whole "is"
the game.

XI

Technologists only change the world in various ways in generalized indifference; the point is to think the world and interpret the changes in its unfathomability, to perceive and experience the difference binding being to the nothing.

Concerning the Experience of the World: On Heidegger

In his brief work *The Thinker as Poet*,[1] Heidegger makes the attempt—without any claim to completeness (a sacrifice forced upon all thought experiments of today) and both through thought and poetically [*denkerisch-dichterisch*]—to give to language (full of intellectual richness although quiescent) an overture or sequel. Heidegger seeks the word, the truth, openness, the meaning of being as be-ing (*Seyn*). "Truth means sheltering that clears [*lichtendes Bergen*] as the primary quality of be-ing [*Seyn*]," he writes in "The Essence of Truth";[2] initially, be-ing already appears "in the light of a withdrawal that conceals," and this sheltering that clears first makes possible and "allows the agreement between knowledge and beings [*Seiendes*] to become present," hence the "correctness of ap-prehension [*Vor-stellen*]." Although Heidegger perceives the prevalence of errance "in the

---

1   EN: *Aus der Erfahrung des Denkens*; "The Thinker as Poet." A literal translation of the title would be "Concerning the Experience of the Poet." Axelos's title for this essay is a clear reference to this.

2   EN: "Vom Wesen der Wahrheit," in GA, vol. 9, 201; "On the Essence of Truth," in *Pathmarks*, 153.

simultaneity between unconcealment and concealment"—for "the concealment of the concealed as well as errance belong to the fundamental essence of truth"—, he subjugates errance, "the pervasion of errance by secrecy," to the truth: "Errance is the playroom [*Spielraum*] of that turning point in which in-sistent existence [*in-sistente Ek-sistenz*] agilely forgets and mistakes itself. The concealment of concealed beings as a whole [*Seiendes*] prevails in the unconcealment of specific beings, as this unconcealment—as the forgetting of concealment—becomes errance."[3]

Is there, however, a highest word for the prevalent and withdrawing, open and fragmentary world play [*Weltspiel*] of the whole of being-and-the-nothing in its becoming?

Since *Being and Time*, Heidegger attempted to render being "as" time. Being is sometimes united with the nothing and experienced as the same. The enigmatic difference between being and beings [*Seiendes*], the duplicity [*Zwiefalt*] within which nothing divides being and beings (for the nothing of beings "is" being and being is not beings), negates both in their simplicity and remains inseparable from Dasein: being and human essence belong together, are not connected with one another; one does not depart from the one only to move to the other, or vice-versa. Sometimes even Heidegger interprets being as "the game itself [*das Spiel selber*]," but this interpretation disappears once again.[4]

Being is not to be saved. And neither its "necessary" playmate, human Dasein. The lightning bolt strikes down into the boredom of time, if the world itself has not been shaken in its very foundations. The world manifests itself as the "mirror-play." Within the play-of-time space in which moving and absolutely motionless time, the past and the having-been, the present

---

3   EN: "Vom Wesen der Wahrheit," in GA, vol. 9, 196; "On the Essence of Truth," in *Pathmarks*, 150.

4   EN: Martin Heidegger, "Identität und Differenz," in GA, vol. 11, 72. The existing English translation by Joan Stambaugh, *Identity and Difference* (1969; Chicago: University of Chicago Press, 2002), 66 is almost unrecognizable.

awaiting us and the already temporalized future convene and
self-destruct.

Distressed and exasperated, we must experience what we know
with insouciance and with excitement. Nowadays, a multi-
facetted and ambiguous, productive and questioning form of
thought seeks its way and its style, internally connected and
intertwined with the attempt to achieve a uniform and multi-
dimensional lifestyle. Thought can no longer be experienced as a
thing of the head—or of ideas, the mind, or be-ing [*Seyn*]. Worldly
expanse and small-world spirit correspond to one another very
often. The planetary horizon, in the four-dimensionality of its
plans, waits for its playing-together—those supporters who can
be neither shepherds of being nor governors of the nothing? In
togetherness there lies a common affiliation within the game.
When each existence is dissolved and restored, when every "is"
becomes fixed and negated, when every word renders a thing
and disintegrates, man throws away his deck of cards: here is too
little of the world rather than openness to the world. Everything
unfolds and converges and disintegrates with the play of time
of errance. Time, world. Errance, game, are they—if they "are"
indeed at all—names of the nameless self-identical?

This sphere of thought is "merely" the whole half of the half
whole. And the other half? Important faces and masks of beings
"and" being will always be neglected. Questions and counter-
questions cross and thwart one another. No problem becomes
solved. The secret itself becomes questionable. The abstract and
the concrete, the positive and the negative—that is, which still
bears this name—merge inseparably. As certain stars perish,
others appear—as well as the self-identical. Individual and universe—both finite, for us, finite beings [*Seiendes*] who question
being, correspond to one another and do not correspond to one
another. Every worldly management and every world plan fails
due to and in the world. The single human and society cross out

and become crossed out. In his work "On the Question of Being,"[5] Heidegger dares to cross out being (which one?) and to demand the same for the nothing. As a sign, this crossing-out points in all four zones, the worldly zones—of the fourfold [*Geviert*], which belong to a unity; earth and heaven, the divine (gods and God), the mortal (humans). The simplicity of this quartet remains intact. Without evading the questionable and elevating the matter to the level of the thought worthy, in his essay dedicated to "Building Dwelling Thinking,"[6] Heidegger makes the attempt to preserve rescue, reception, expectation, and dwelling. But how can all of this—the one-all—persist intact, if a "new" experience of the world is dawning? Beings and undergoes destruction, that which has been crossed out, remains as the sign of crossing, pointing to the perspectives of the worldly fourfold, as an unknown x.

Heidegger places in question the Occidental and European interpretation of being handed down by world history, which simultaneously means the oblivion of being [*Seinsvergessenheit*]— from Heraclitus to Nietzsche and Marx.[7] He inquires and gives inquisitive answers. His attempt demands to be placed in question itself.

> Discover and conceal,
> cheerful word and dreary lie
> lie upon a single path.
>
> Declare and deny
> err and dare
> askew to the same and other bridge.

5   EN: Heidegger, "Zur Seinsfrage," in GA, vol. 9, 411; "On the Question of Being," in *Pathmarks*, 310.
6   EN: "Bauen Wohnen Denken," in GA, vol. 7, 145–64; "Building Dwelling Thinking," in *Poetry, Language, Thought*, 145–61.
7   See Kostas Axelos, *Héraclite et la philosophie. La première saisie de l'être en devenir de la totalité*, Collection "Arguments" (Paris: Éditions de Minuit, 1962).

The world is shaken in its very foundations and errors uncover the systemic fragments of a game whose beginning and end remain in circular contortions and in concealment.

When gods die, man begins to dwell with the anticipation of his own death. In the prosaic poem about the overcoming of man just begun.

Consent to falling stars.

Thought is entry into the labyrinth of the world, which does not obey a highest name.

When smoke from the chimney betrays the ashes . . .

When the courage of thought is in harmony with the celestial hours of destiny, it gives expression to the resting course of the mad stars.

Since ancient times, seeing and hearing have made thought attentive and unhappy.

The coming, generalized combinatorics put everything into relation with everything else, generate new divisions, and release powers of the universe.

Every conversation brings together human creatures who are mute and deaf.

When bright trails become visible against a dark sky . . .

Who or what generates thought?

Dialogue is always an interlude [Zwischenspiel].

Mortals cannot fight hard for or against experience of the world. Success overcomes players and spectators and disintegrates. The wind encourages and batters to pieces.

Both in performance and in failure, dexterity and attempt are at risk. Masters always become mastered.

When the flowers of the meadow deceive us like a fairytale . . .

The manifold nature of the simple.

Image and countenance are changed through their masks, into concrete abstraction.

Hope and hopelessness are borne by one another.

The cunning of pain is probably more widespread in the world than that of pleasure.

When the wind alarms everything weak and steadfast . . .

Three unified games play [Spiele spielen] thought and world.

The world game itself, which is placed under regulations.

The game of thought that constantly derails on examples. It must dissolve every existence, dwell clumsily and without boundaries, which it is only seldom able to do.

The game of humanity: fancying danger has been avoided when man desires to be the player.

When a locality opens and closes: like a wound and a flower . . .

In the voice of silence, enunciating thought seems to unfold as the whole half of the half whole.

How can one distinguish between thoughts and things?

A long while and a short term act out and squander themselves within the circle of time.

Whoever thinks great things, must they live as a Philistine?

When jagged boulders grip cunningly with their gargantuan irony
. . .

The old and the new appear erroneously as two sides of the same coin.

For this reason, comprehend foreplay as unfolding and simplification.

To be ripe means: to perceive and experience every place and moment, the present larvae of past and future time, as nearly inconceivable.

Desire and anticipation are one and at the same time different.

When the mountain conceals and stillness becomes noisy . . .

Everything emerges from stillness, returns back to it. In the interim, the ineffable and the unsaid befall us, the nameless and the unthought-of.

Doesn't the unconditional remain chained to conditions and things?

What withdraws: things, words, language, thought, the withdrawal, the world?

That thought can never interpret the world game—whose success would fathom this?

When the shepherds are driven into the pastures by their herds, alluring and lured . . .

The game quality of thought as yet remains concealed.

Where it shows itself, it is similar—in time—to a utopia, which promises and ruins an always inappropriate visit.

But how would a chronology and topology of being and the nothing within the game of time and space [Zeit-Raum-Spiel] be able to persist and decline?

Acknowledging that essence [Wesen] is also mischief [Unwesen].

When the evening light can no longer be distinguished from the light of dawn . . .

To say and to think, to compose and to act are the neighboring branches of a puzzling tree.

They arise from the world and almost reach down into the root of errance.

Their concord provides the experience of what Novalis stutteringly said about their common providence:

"Truth is a complete error."[8]

> Question aims
> Fracture questions
> Stars rest.
>
> Man waits
> Games aim
> Without why.
>
> Riders fall
> Children hope
> A sound deed.

---

8   EN: Novalis, *Werke, Tagebücher und Briefe Friedrich von Hardenbergs*, ed. Hans-Joachim Mähl and Richard Samuel, vol. 2 (Munich: Hanser, 1978), 449.

III: THE PLANETARY

The Planetary: A World History of Technology

> *But let nobody claim that fate divides us! It is us, us! We have our pleasure when we plunge into the night of the unknown, into the cold strangeness of some other world, and—if it were possible—we would abandon the sphere of the sun and storm beyond the limits of that mad star. Alas! For the wild breast of man, no homeland is possible.* (Hölderlin, *Hyperion or the Hermit in Greece*[1])

I

Let us attempt with complete sobriety, plainly and flatly, to take into account how what we still call a world can be taken into account, and how we—on the threshold to the planetary

---

1   EN: Friedrich Hölderlin, "Hyperion oder Der Eremit in Griechenland," in *Sämtliche Werke und Briefe in drei Bänden*, vol. 2, 24; "Hyperion, or the Hermit in Greece," in *Hyperion and Selected Poems*, ed. Eric L. Santner (New York: Continuum, 1990), 10.

age—can intervene technically, *planning* and *planing* [planmäßig *und* planierend], in the whole of a world split into pieces.[2]

Let us assume that beings as a whole [*das Seiende im Ganzen*] constitutes the "totality" (the uncanny homeland of homeless modern humans),[3] the entire realm [*Gesamtbereich*] of all experience. This whole no longer continues to appear as a unified all. The united and uniting destiny, which perhaps still pervades a delimited circle, cannot show itself. The all itself, and everything that exists, appears fundamentally divided and split into different regions, sectors, districts, and layers, to which different points of reference and perspectives are supposed to correspond, and it is these perspectives that delimit those areas and layers. This great, crude division of totality affects the difference between *nature* and *history*, although the only shared foundation of both "parts" remains unperceived and unconsidered.

Nature is considered to be the entire realm of everything that grows (*phúetai*) and shows itself (*phaínetai*) without having been produced by humans. The *physis* and *kosmos* named by the Greeks, which was reinterpreted by the Romans as *natura* and *universum*, transformed itself, i.e., became transformed into "nature" during the course of a long history. Thus, nature is viewed as "the sum-total [*Inbegriff*] of all things, to the extent they can be the objects of our senses, and correspondingly, of our experience" (Kant).[4] The planetary system, the earth and the sun, water, air, fire, matter and energy, rock formations, flora and fauna, and finally humans, as natural living beings, belong

2 EN: Axelos's terms are hard to capture in English. We have opted for planning—scheduling, organising—and planing—flattening or levelling.

3 EN: The German is *die unheimliche Heimat des heimatlosen modernen Menschen*—the unhomely homeland of homeless modern humans.

4 EN: Immanuel Kant, "Metaphysiche Anfangsgründe der Naturwissenschaft," in *Kants Gesammelte Schriften*, vol. 4 (Berlin: Georg Reimer, 1903), 467; "Metaphysical Foundations of Natural Science," in *Theoretical Philosophy After 1781*, ed. Henry Allison and Peter Heath (Cambridge: Cambridge University Press, 2002), 183.

to the powers of nature. The natural sciences that have grown together with technology investigate and process nature systematically—*physis* which has become the object of physics—, and each of its sectors and forces has an adequate knowledge of practical application corresponding to it. Cosmology, astronomy, mechanics, chemistry, geology, botany, zoology, and biology scrutinize the entire hierarchy and genesis of beings in nature. This hierarchy, which the Greeks knew quite well without, however, making a rigid schema out of it, was constructed methodically and edifyingly in the *Book of Genesis* of the Old Testament, and still dominates worldly interpretations of the sciences bearing the name developmental theory. The human is considered to be the culmination of this evolution.

History begins with the last creation of the architect of the world. The worldly events caused by the human species comprise a primitive "prehistory" still bound to nature in an original and archaic way, which the so-called "indigenous peoples" of today can still convey to us in a particular—although uncertain—manner. These happenings become a "real" history after the Oriental and East Asian empires: the cradle of civilization in Mesopotamia, Egypt, India, China, Palestine realize the transition from nature to the spirit of culture. However, actual history [*eigentliche Geschichte*] begins in the Occident, i.e., in Greece; here, for the first time, the truth of destiny emerges. In Greece, the powers of language and thought, of literature and art, of religion and politics unfold, form a holistic unity, which still remains the foundation of all education—including planetary knowledge. The pre-Socratics who apprehend the essence of *physis* and give expression to *logos*, Socrates and Plato who establish the hegemony of the supernatural and non-sensual *idea* over the sensual in their battle with the sophists, Aristotle—the thinker of *energeia* and *being-ness* (*ousia*), the three abrogated schools of thought stoicism, Epicureanism, and skepticism, and finally, the encounter between heathen thought and the religious faith of Plotinus and the neo-Platonists (who still seek the deceased one), constitute the main phases and the internal intellectual evolution of Greek history.

Greece declines and Rome rises—the less thoughtful than proactive Rome; the realistic, republican, and imperialistic Rome with its *republic* and its imperial rule, with its history-making laws. Greece and Rome execute the first authentic performance of world-historical proportions and founded classical antiquity as a result.

The second great step completed by Jews and Christians leads to the Reformation. Its climax takes place during the Christian Middle Ages. The Old Testament and the Jewish prophets, the Gospels and the New Testament, the Church Fathers, Augustine, mysticism and Scholasticism provide the foundations for the prevailing faith of this epoch, as they process and solidify the belief in the Biblical revelation of God who became human and died. Everything that exists appears to be a creation of the Creator and is subordinate to divine providence. Everything that becomes is a product of the divine *actus purus*, which every human action should correspond to.

The decisive historical step leading to the planetary era, however, is only the third: the modern European epoch. The modern human enters the scene, the subject that will dominate all objects with its thought and knowledge, action and influence. This third epoch desires to be a reincarnation, a renaissance, and cannot remain merely European for all that long: it moves along a track leading to a premeditated and consummated history of the world that encompasses all parts of the earth. An uncanny power drive compels the epoch of the incipient will to power. Descartes thinks the *ego cogito*, i.e., *res cogitans* as the objective subject which stands opposed to objects within the *res extensa*; he is the founder of the logic of rational intervention. Pascal conceives of the reason of introspection, the *raison* of the heart. Spinoza attempts to grasp all-encompassing, natural and divine substance. This substance manifests itself with two attributes, however: extension (matter) and thinking (spirit); and his *Ethics* has a mathematical and geometrical structure. Leibniz deals intensively with the question: *pourquoi il y a plutôt quelque chose que rien?* [why is there something rather than nothing?]—and sees the answer in the general direction of the: *ens percipiens et appetens.* The subject is the *ens* of beings [*Seiendes*]; its perception and pursuits provide the foundations for the world order. *Metaphysically* and epistemologically, Kant establishes the power of the transcendental subject who usurps the existence of beings as its objects. The transcendence of substantiality (objectivity) and the transcendence of subjectivity (of non-isolated solipsism) subsist in the same reality. Metaphysically, Hegel ends an entire epoch: spirit that has become nature and human history manifests itself as absolute knowledge, which consummates the truth of the whole through absolute self-consciousness in the form of absolute certainty. Marx initiates the countermovement: he converts the practical human with its technical activities and effects into the objective subject of concrete reality. Nietzsche draws the conclusion: the rule of the will to power leads to the murder of God, to the age of nihilism, i.e., the destruction of supernatural meaning that had provided hold up to then. Will the *Übermensch* be able to truthfully apply the will to power in order to achieve the planetary dominance of earth? Will humanity that adopts, but also relinquishes and overcomes the human of yesterday—within the planetary rotation of the eternal recurrence of the same will to power—be able to say "yes" to this? Descartes and Pascal, Spinoza and Leibniz, Kant, Hegel, Marx, and Nietzsche are not just any profound philosophers; they are *the thinkers* who initiate present and future events.

But what should the present phase of Occidental and European developments in the modern era be called? The history leading to the planetary stage of evolution does not know all that well how it should be described: as the first epoch of actually realized and unified world history? The nuclear age? The planetary epoch?

The human, its "being," and its historical evolution have been commandeered by an entire army of historians and scholars of the humanities. Can knowledge that has been conquered in this way provide answers to these urgent questions? The human body and the human soul constitute the objects of biology and psychology, whereas they are placed under special protection by medicine and psychotherapy. The so-called indigenous peoples are rummaged through by ethnologists. The economy, society, social constructs, and politics give national economies, sociology, and the social sciences a great deal to do. The forces that unfold during the course of historical events—religion, literature, and art—have also become objects of the proficient humanities. And, finally, the entire historical process, as well as everything that occurs in and through it, becomes assimilated and elucidated by historical science, i.e., history.

Thus, beings—whether they be a natural entity or a product of human-history—find their treatment and classification in accordance with the natural sciences and humanities, within a total plan and reconnaissance. An at least two-fold treatment is assigned to everything that has become the object of knowledge and to every matter of concern: a systematic and scientific (which is intended to apprehend the matter) and a chronological and historical (which investigates its emergence and development).

For its part, the most distinguished deed of human performance, thinking, is taken over by philosophy, and philosophy—as metaphysics—analyzes the being of beings as a whole [*das Sein des Seienden im Ganzen*]: as idea, God, spirit, human, matter, energy, enframing [*Gestell*], or according to whatever interpretation, and finally announces its judgment on natural events without faltering, while striving to construe human history. Accordingly, this philosophically comprehensive thought becomes even further subdivided to a significant extent. Metaphysics—*metaphysica generalis*—, ontology, is directed at the whole, at the entity as entity. Logic is the judge of thought, and epistemology remains equally determined by the *ratio*. The components of a developed and dissolved *metaphysica specialis* are concerned with regional ontological zones. In this way, natural philosophy, anthropology, historical science, ethics, and aesthetics exhaust nature and its powers, the power and powerlessness of historical man, the virtue of action and the beauty of artworks. Philosophical and scientific history, i.e., the history of philosophy, finally investigates the historical process of doing philosophy and its respective stations.

Philosophy, i.e., (open) thinking which has become philosophy, already underwent subdivision and gradation into "logic," "physics," and "ethics" under the first philosophical academy. Beginning with ingenious Platonic school philosophy, strongly colored and systematically transmitted by the theological exegesis of faith in Biblical revelation, this plan leads to practical action in diametrical opposition to theoretical thought, all the way to Hegel. *One* ontological schema, subdivided into regional ontological zones and subject to the power of methodological perspectives, circumscribes the being of beings as a whole as well as the regions of totality.

The *logos* of the whole, its being and its truth, the dialectic of events (interpreted as idea, God, spirit or meaning, and the direction of moving, energized material), is apprehended "once again" by logical human thought, and becomes articulated, conscious of itself through human language and human consciousness. As a result, logic and dialectic correspond to the *logos* of being or becoming.

The realm of nature, its matrix of order, its development, and its entire hierarchy have been placed under the focus of physics. After *logos* or spirit has illuminated everything, the massive naturalness of beings can be taken into account and adopted as a task by the suitable branches of knowledge and technical activities.

Through the human, nature is transformed into something different. Through the human and the spirit of human history, all empirical events enter the locus of self-consciousness. This attainment of certainty is the foundation of ethics and is borne by the latter. Humanity brings a total plan to fruition, plans systematically and levels that which offers resistance. Thus, we achieve our objective: the planetary age of the earth, this wandering star.

Everything is so perfectly in order and clear, plain and simple. There are beings [*Es gibt Seiendes*]. Metaphysics and speculative thought question the being [*Sein*] of beings, the living or dead God, and expose—spiritually or materialistically—the core of everything that exists. The being of beings is considered to be the *logos* of all that occurs, as a general and total plan with which human thought is said to exist in harmony. Initially, beings only appear within nature, which somehow began and proceeded to generate mankind: the natural sciences rush towards physical nature and penetrate it. The history of nature—and human creatures—fills the second great region of the totality. History begins with prehistory and finds no peace until it arrives at the consummation of planetary world history. During the course of this history, humanity learns to speak and think, the powers of religion, literature, politics, philosophy, science, and technology unfold, and they give expression to the whole and its various components, while they themselves can become the objects of varying perspectives.

The circle closes, and we of today investigate the being of beings, do work in the natural sciences, have comprehensive historical sciences at our disposal that encompass the whole of our efforts, while our thought somehow manages to apprehend everything adequately and at the right place. All the while, our technical activities intervene everywhere in order to alter things systematically and pragmatically. In this way, the history of the world is realized in a uniform and total manner as world history, all humans and peoples of the earth think according to the same plan, endeavor the same things, and are driven on by the same things, all are of the same breed. A variegated but focused form of thought, an efficacious form of science, and total technology grip the entire planet and lead unified, i.e., uniformed humanity on to the radical domination of nature, to the satisfaction of all natural (and even intellectual) needs and drives. Collective humanity presents itself as *the* maker of plans; it is the objective subject that grasps all concrete objects in order to "actually" transform them with its practical and technical will that knows no bounds.

This total and unitary construct is present today in all countries and in all minds; its variations are part of its nature. Jews and Christians, positivists and idealists, citizens and Marxists pursue the same thing, whether they are aware of it or not. Everywhere everything is taught and learned, done and produced in accordance with this schema. The same plan is followed across the global surface of the earth: in a technical and theoretical as well as technical and practical mode.

All beings—be it an apple, the smile of a child, or a lathe—are taken into account in a mode compliant with technology and the natural sciences, pored through in accordance with historical criteria, and reshaped in a practical manner. Everything which has being is also interpreted in terms of mythology, the history of religion and the humanities, and all forms of manifestation within mythology, religion, literature, and the fine arts are subjected to efficient analysis. In the end, everything that has being will eventually be rendered as a word, concept, or sign by linguistics, logic, epistemology, and logistics, while metaphysics with its speculative thought prepares itself to tirelessly question the being of this being. The same holds true for every single being— be it an apple, the smile of a child, or a lathe.

After a long prehistory, we have finally entered an age that has generalized beings [*Seiendes*], investigated them in a multifarious and comprehensive manner, and actually, i.e. technically restructured them. This age should be called planetary.

Is not everything in order this way, and is not this order itself the planning and planing planetary world order which is according to plan? Can a fracture be seen somewhere beneath the shining Sun-star, which has not been considered a planet for several centuries now, as well as on the earthly wandering star [*irdischen Irrstern*], which has only become a planet since the Copernican Revolution and then plunged into splendid rotational motion?

II

Why, however, should this present age be called planetary? Because its orbit is a transformative and wandering path? Or because it is an errant path? Because its plans are planetary, or because it plans to level off the universal [*das All*] during the course of its journey?[5] Because the order of its plans leads to total planification and—in spite of this and due to this—the age of a wandering star that reveals itself as a wandering star comes into existence and remains so? Because everything is stretched over a faceplate [*Planscheibe*] whose transforming rotation persists with the same indifference? Or because we cannot envision this final plan of the entire picture and enframing [*Gebilds und Gestell*] and give adequate expression to them?

According to the Greeks, the essence of the "planetary" lies in a wandering that has gone errant; *plázo* (future) *plágxomai* (aor. pass.) *heplágkthen* (Latin: *plango*), an authentically Homeric term, means: *to strike* and *to be struck* and *to be driven, to err, to plague, to wander aimlessly.* The full meaning of the Odyssean adventure, and not only of this adventure, is alluded to in the opening verses of that epic quest: *O Muse, tell me of the deeds of that many-sided man, / who journeyed off so far after the destruction of sacred Troy, / who saw and learned the cities and ways of many peoples, / and on*

---

5   EN: Das All is not simply translated. It can mean the same as the English "all," and does when Axelos talks of the "all of beings," but in this context implies something closer to "cosmos" or "universe." We have used "universal" to try to capture the two senses.

and the return of his companions.[6]

Human beings are struck and driven on by the being of *physis*, by
destiny, by the lightning bolts of Zeus, *plazómenoi*, and they are
continually one with wandering, erring individuals on an odyssey.
They are *plánetes*: they are the errant ones. Mortals roam and
go astray on earth, which is not a wandering star for the Greeks.
They err, but not in respect to a correct truth. The *aletheia* reveals
itself during this journey, and man errs and wanders within the
clearing (of unconcealment). "The flight of the human away from
the secret to the practicable, from the commonplace, to the next
thing, past the secret, is *erring*. The human errs. The human does
not at first enter errance. But only they ever go astray. . ."[7] Erring
should not be understood as incorrectness, as falsehood, or as
making a mistake, as deviation: we "err" quite crudely, if we do
so. "Not any isolated mistake, but rather the monarchy (the rule)
of history by that inherently conspiratorial involvement of all
forms of erring is the error. [. . .] The errance, in which historical
humanity must respectively go astray in order to make its development errant, plays an essential role in summoning the openness of Dasein."[8] The Greek terms *plazo heplagzthen*, *plazómenoi*,
plánetes, *planetai,* name the event of wandering and going astray,
and thus, also give expression to errance as mistakenness (*pláne*);
we ought not, however, the latter using the powers of the faculty
of judgment [*Urteilskraft*].[9] "The disclosure [*Entbergung*] of beings
as such is at the same time in itself the concealment of beings as
a whole. In the simultaneity of unconcealment and concealment,

6   EN: Homer, *The Odyssey*, 1, 1–5.
7   Martin Heidegger, *Vom Wesen der Wahrheit*, 2nd ed. (1943; Frankfurt am Main:
    Vittorio Klostermann, 1949), 22. EN: "Vom Wesen der Wahrheit," in GA, vol. 9,
    196; "On the Essence of Truth," in *Pathmarks*, 150.
8   Heidegger. *Vom Wesen der Wahrheit*. 23. EN: "Vom Wesen der Wahrheit," in
    GA, vol. 9, 197; "On the Essence of Truth," 150–51.
9   EN: Kant's third critique was entitled the *Kritik der Urteilskraft*, the *Critique of
    Judgment*, or, more literally, of the *Power, Force or Faculty of Judgment.*

errance pervades. The concealment of the concealed and errance belong to the original essence of truth."[10]

In the view of the Greeks, the planets are the wandering stars, the mad stars, the driven stars, the *plánetes* (*asteres*). The Greeks have neither experienced for the first time the essence of errance through the motion of the stars, nor have they transferred this essence from the human to the cosmic level. The fundamental experience of errance revealed itself to them but remained nonetheless withdrawn. The sun itself, the shining star that leads to unconcealment [*Unverborgenheit*], is a "planet" for them.

"(God) guides everything that crawls with blows."[11] This divine castigation (*plegé*), is "the lightning bolt that guides everything."[12] Everything that crawls and wanders across the level plane is *driven*, *struck*, and *guided*.[13]

The unity of the all of beings, the prevailing *physis*, bears truth within it, which humans—through their errant wandering—unconceal, e.g., generate through speech (through *logos*), through production (through *poiesis*), through action (through *praxis*). Thus, this wandering and going astray of mortals occurs in one place: the one-all (the Heraclitean *hen pánta*) is this true place.

---

10   Heidegger, *Vom Wesen der Wahrheit*, 23. EN: "Vom Wesen der Wahrheit," in GA, vol. 9, 197; "On the Essence of Truth," 151.

11   Heraclitus, frag. 11. EN: Here and below, Axelos cites Heraclitus by the standard Diels-Kranz B numbers, which enable reference to multiple editions. For English translations, see "Heraclitus," in *Early Greek Philosophy*, trans. and ed. Jonathan Barnes (London: Penguin, 1987), 119.

12   Heraclitus, frag. 64. EN: "Heraclitus," 104.

13   The determinative factor in the words: *plazo* (strike), *plánes* (wandering and gone astray), *plánétes* (planet), *planus* (plain and level), and *plânus* (roaming) is not their shared nor their divergent etymological origins. Several linguistic hypotheses make the attempt to explain these implications—by approaching their essential meaning (*pelázo*) and withdrawing from it. Readers can find these differences in connotation easily enough by in the appropriate dictionaries; however, the common meaning that is expressed—and the matter at hand it expresses—is much more difficult to conceptualize.

Technê and *physis* are and remain intrinsically interconnected for the Greeks, and even possess a common and single origin. However, the mystery of their birthplace remains unsolved. Natural occurrence and poetic production, cosmic collapse and active influence (*praxis*) are determined by a sameness, although this sameness remains mysterious, and we cannot comprehend how something that has split asunder so violently could have been so intimately united at the beginning.

The emergence of the Judaeo-Christian belief in revelation bestows new luster on beings as a whole. God is the being of beings, the absolute forger of plans, and the plan of providence encompasses the entire genesis of the world: from the initial act of *creatio ex nihilo* to the apocalypse. The logos of God generates all of creation according to a systematic plan—of nature and humanity, and mortal sinners must—battling against nature according to plan—subjugate themselves to divine providence. Accordingly, humans develop the skills to complete this divine plan, and everything is reflected in the plane mirror of fundamental belief. God who became human shows humanity the only way. Through this belief in revelation—as well as its probation and petrification by authorities of the church, everything that appears as "natural," "material," "corporeal," and "sensual" becomes part of this battle, whose final objective remains a supernatural realm. Godlike—but fallen, sinful, and guilty—the human must dominate and transform everything through his own actions; the human has become the second greatest lord of creation; divine creativity has been taken over by godlike human production.

For us humans of the modern era, the earth-globe [*Erdball*] has become a wandering star. The history of humanity is unfolding across the entire surface of the earth-sphere [*Erdkugel*], enveloping it and the surrounding space. Human plans level anything that opposes this fate. We are the first planetary age of world time. A total plan takes hold of everything and ignites all human drives, while earth itself becomes the battlefield of these designs. Everything is meant to happen according to plan in order to achieve a gradation that is total and world-historical. With the help of technology, everything is stretched upon a faceplate whose rotation is presumed to parallel the rotation of the earth-sphere. As we are the ones struck and guided by every power, however, we have "lost" the place for the revelation of truth.

Do we err, are we errant, or have we gone astray? The essence of the Greeks arose from one foundation and inspired one place; its errant wandering took place within the truth of beings as a whole [*Seiendes im Ganzen*]. The essence of Christianity also ignited a true locus through its erring. But we, on the contrary, do not have the truth: "The new aspect of our current position on philosophy [and not only on philosophy] is the conviction that no age has ever had before us: *that we do not possess the truth.* All of previous humanity 'possessed the truth,' even the skeptics."[14] The question remains open: perhaps the truth possesses us. In any case, truth has ceased to illuminate a singular place and has been drawn into wandering errance; it remains concealed within the circular motion of the planetary time-space.

The age before which and in which we stand (i.e., staggering and roaming), the present course of worldly time that has already dawned (although it has not yet truly begun), is *planetary*: planning and planing all that exists, placing it on the faceplate according to plan, consummating a total plan. A planned economy and the stubborn economic struggle leading to planification con-stitute merely an extremely visible façade as well as one of the effective powers within this holistic plan of gradation; although economic operations are intended to lead to the satisfaction of all

---

14   Nietzsche, W. W. XI, 268. EN: This is a reference to the *Großoktavausgabe* collection of Nietzsche's works (ed. Elizabeth Förster-Nietzsche et. al., Leipzig: Kröner, 1894–1904) though this has been long been superseded. The passage in question can be found in KSA, vol. 9, 52 [Notebook 3 (19), from early 1880]. There is, to my knowledge, no English translation of this passage.

drives and needs, and is also viewed as an impetus (in Marxism),
it still remains subservient to the power drive.

The roots of this rootless age that is attempting to consummate
the destiny of technology and nihilism lie in the objective positing
of human subjectivity (as *res cogitans*), in the planned attack
on the entirety of *res extensa* (objective and effected reality),
in the collapse of perspective and in combinatorics, in the will
to know and the will to dominate. There "is" no more meaning
of being, being has become an errant and wandering genesis,
and everything that is has become the object of a planetary
technology according to plan, which grasps violently into this
emptiness. The destiny of the world—i.e., of the openness of
"becoming" being, turns upon world history, and the world—i.e.,
the non-world—appears as a work-world [*Werkwelt*] within which
no space can remain for the true tasks of existence.

The destruction of the—merely extrasensory?—truth of being of
the whole of being in its genesis [*werdendes Seiendes im Ganzen*],
the "negation" of the meaningful openness of the world, and an
active nihilism that only places sensual beings on its faceplate,
achieve an enormous accomplishment and execute a deed
of world-historical proportions: the power compelling them
conquers the visible whole of the erring planet earth, unites
humanity, unconceals all that exists by artificially elevating it
to the object of planned production. In this way, something
fundamentally new is created. This entire machination, this
business of fabrication, however, possesses neither truth nor
meaning, neither a place nor answers to the questions: Why? To
where? For what? Despite and because of this, it does not stop
by any means, but continues on and on; the world wars are an
essential part of this scenario, although they—perhaps—make
the aggressive technology employed there only superfluous.

It cannot be predicted at all whether this planetary process
will lead to world catastrophe, to world destruction, to world
decline. Perhaps it is the most profound foundation and final

ground of beings to perish "in the end." Perhaps. It is easy to state; difficult to conceive of. World destruction? Why not? Why not nothing? That could be a response to the question *Why?* by means of an (absent) answer. "We are making an attempt with the truth! Perhaps humanity will perish as a result! All right then!"[15] wrote and cried Nietzsche. The question posed by philosophical metaphysical thought since ancient times, the question *tí tò ón*, the question posed by Leibniz: *pourquoi il a plutôt quelque chose que rien?*, the message that can be expressed in a query by asking: *Why are there beings at all and not, much rather, nothing?*,[16] being itself, which has become so questionable, will all of this and its world-being perish? And should this nothing be preconceived as fullness or emptiness? The speaking question is already difficult to hear; but it is even more difficult to pronounce the words of the answer.

The history of modern European thought that primarily takes place within metaphysics, i.e., philosophy—will it at least help us find the way? The event of modern Occidental thought leading to the planetary age—will it illuminate the course of the planet and where does it receive its light from? Can we calmly continue along our mistaken path? In any event, this thought does not execute sundry and varying steps. The power of this thought unfolds in a few strictly regimented steps leading to the realization of metaphysics, to the "becoming philosophical of the world" as the "becoming world of philosophy," provided that this realization is at the same time its loss.[17]

15   Nietzsche, W. W. XII, 307. EN: KSA, vol. 11, 88 [Notebook 25 (305), from early 1884]. No known English translation.

16   Heidegger, Martin, *Einführung in die Metaphysik* (Tübingen: Niemeyer, 1953), 1 and *Was ist Metaphysik?*, 20. EN: GA, vol. 40, *Einführung in die Metaphysik*, 3; *Introduction to Metaphysics*, trans. Gregory Fried and Richard Polt (New Haven: Yale University Press, 2000), 1; "Einleitung zu 'Was ist Metaphysik?,'" in GA, vol. 9, 381; "Introduction to 'What is Metaphysics?,'" in *Pathmarks*, 289.

17   EN: Marx, "Aus der Doktordissertation," 17; "Notes to the doctoral Dissertation," 62.

Descartes completes the first decisive step: he posits the subject, the thinking and acting *ego*, the *res cogitans*; this subject is
argued to dominate the entire reality of the *res extensa* through
its self-reflection and effective action in order to make rational
use of it. Since first finding expression through *logos*, physics
and metaphysics were intimately interconnected in an unusual
way. Now, metaphysics is beginning to find its culmination in
physics and this can occur due to the fact that physics originates
from metaphysics, whereas metaphysics is strongly influenced
by "physics." As a result, *physics—as metaphysics—now converts
physics to technology.* The human subject, who desires to gain
power over objects by means of rationality, is itself objective, i.e.,
posited as an object and set on its way.

Kant takes the same route: he posits the transcendental ego and
attempts to establish its foundations. This thinking and acting
ego apprehends objects as objects [*Gegenstände als Gegenstände*]
of its experience, i.e., as objects [*Objekte*]. The transcendental
quality of objectivity subsumes transcendental subjectivity and
simultaneously gains its foundations from the latter. The transcendental subject and transcendental object exist in a necessarily relation to one another and have their foundations in the
same.

Hegel strives to illuminate the meaning of the entire process: he
interprets the total dialectical development of mind leading from
the logos (of God) to the naive realm of nature and onwards to
the spirit of human history, whereby the process as a whole is
apprehended in the end as absolute knowledge by the subject,
by self-consciousness. Truth which evolves in this manner is the
truth of the whole—"that which is true is the whole"—, and the
dialectic of the thinking subject gives expression to dialectic as
the momentum of becoming being, i.e., as a dialectic of concrete
and objective reality itself. Much rather, idea and reality are the
same.

Something reaches its conclusion with Hegel; that having been said, we do not yet know precisely what culminates with him and presages a new future.[18]

The posited objective subject—of human being—that desires to dominate all objects (the whole of beings as *the* object) and transform them at the practical level through its intellectual faculties, its effective will, and its productive labor—begins to run its path once it has been set in motion. The subsequent stretch of the path is to be traversed by the acting and generalized

18 Beginning and end, rise and fall, that which is past, present, and future, the old and the new, tradition and revolution disguise each other mutually, hide one another amongst themselves, and one after the other. Each of these powers contains its "contrary pole" in a certain sense, presupposes it, and is "explained" by it. They move in circles. Where and how does something begin, how does something stop—or find an end or its own particular end, how and when does something new emerge? What is transformed and how does something new come about? The transitions can only be explained with difficulty. Marx comments quite soberly on the beginnings of the new society: "a communist society, not as it has developed from its own foundations, but on the contrary, as it has directly *emerged* from capitalist society; which remains bound in every respect—economically, morally, intellectually—to the old society from whence it came." (*Kritik des Gothaer Programms* (1875; Berlin: Neuer Weg, 1946), I, 3 [EN: "Critique of the Gotha Programme," in *The First International and After: Political Writings Volume 3*, ed. David Fernbach (London: Penguin in association with New Left Review, 1974), 346]). At the same time, however, as we shall see, he wants to believe in an entirely new future: for: "communism is different from all previous movements." Heidegger claims that the present age "begins nothing new, but only carries to an extreme the old which was already presaged in the modern age." (*Unterwegs zur Sprache* (Pfullingen: G. Neske, 1959), 265 [EN: GA, vol. 12, *Unterwegs zur Sprache*, 253; *On the Way to Language*, trans. Peter D. Hertz (San Francisco: Harper, 1971), 133.]) At the same time, he asks the question: "*Are we* the latecomers that we seem to be? Or are we not at the same time the early arrivers in the early stage of a completely different period of world history, which has let us abandon our previous conceptions of history?" (*Holzwege*, 300–1 [EN: GA, vol. 5, 326; "Anaximander's Saying," in *Off the Beaten Track*, 245–46]). We exist much more in an *interlude* than in the game itself. Compare as well: Kostas Axelos, *Vers la Pensée planétaire: Le devenir-pensée du monde et le devenir-monde de la pensée*, Collection "Arguments," (Paris: Éditions de Minuit, 1964), primarily the concluding text "L'interlude."

objective subject. Marx intends to place Hegelian dialectic on its feet—for the task at hand is being able to run faster and faster; technological means of production will in fact be those forces that develop most quickly, and they do not lose the real ground beneath their feet. Accordingly, the technically active human is elevated and bestowed the dignity of the objective subject, whereas labor becomes the exceptional modus through which reality is manifested. "Therefore, the greatness of Hegel's *Phenomenology* and its final result—dialectic as negativity, impetus and generative principle—is that Hegel comprehends the self-generation of the human as a process, objectification as loss of objectivity, as alienation, and as the supersession of this alienation; hence, the fact that he comprehends the essence of *labor* and grasps the objective human, the true, i.e. empirical human as the result *of their own labor*," are the words of the young Marx.[19] Nevertheless, Marx does indeed sharply criticize Hegel's conception of labor: for Hegel, labor remains an activity of the spirit—in the opinion of Marx; in his own view, by contrast, labor is the concrete processing of an objective substance carried out by the human subject. The essence of the empirical (reality) consists in objective actualization by means of activated potentiality.

In this context, however, technically active man is not viewed as the individual human. The subject becomes socialized and generalized. It is argued that humanity itself becomes the objective subject through the revolutionary negation of the individual and the private in order to—as an objective subject— subject itself to everything. Marx declares war on philosophical thought, literary and artistic activity, religious faith, and the act of founding the state. All of these activities are criticized as ideology, as idealistic and spiritualistic superstructure, as alienated theoretical performance, as the sublimation of the tragedy of

19   Marx, "Nationalökonomie und Philosophie," 269. EN: "Economic and Philosophical Manuscripts," 385–86.

empirical being and empirical events. A demand is made for the full acknowledgment of the natural, i.e., social essence of man (the subject of objective history), the acknowledgment and complete realization of his objective essential powers. The alienation and divestiture, within which the entire previous history of humanity—and especially modern European history—has developed, should be ablated through the total and planned release (liberation) of productive powers, through the actualization of the potentialities of technology. Everything that exists becomes the material of labor, and collective, social, and socialistic humanity is the objective subject of this concrete and absolute, technical and productive praxis that never ceases to generate new objects. Marx clearly stated "communism is a highly practical movement that pursues practical objectives with practical means, and which is able to—above all in Germany and in respect to German philosophers—take a moment to tend to 'the essence.'"[20] The effective essence of communism, the battle which must be fought by technology against every organic entity, was defined by Marx himself: "Communism is distinct from all previous movements in that it overturns the foundations of all previous relations of production and social intercourse, and for the first time consciously treats all organic prerequisites as the creations of pre-existing humans, while ridding these prerequisites of their organic nature and subordinating them to the power of united individuals. Therefore, its establishment is essentially economic . . ."[21] The center of Marxian thought, as well as the activity it engenders, consists in the generalization of the subject and the positing of human society as the place and subject of historical genesis. This human society is both a product of technology as well as the technically productive subject itself. Technology (in the form of labor, material production, practical fabrication, concrete generation) is the power that transformed nature into history, and constitutes the innermost driving force

---

20   Marx, *Die deutsche Ideologie*, 218. EN: *The German Ideology*, 231.
21   Marx, *Die deutsche Ideologie*, 70–71. EN: *The German Ideology*, 86.

of world history, which can first become actualized world history during the planetary age. The naturalism-humanism-communism manifested in reality is based upon man, his natural and social drives and needs; it negates all natural quality and it alone is capable of developing objective powers of production completely and according to plan. The driving truth and motivational power of Marxism lies in its comprehension of planetary technology, of world history as the history of objective action, of the being of the world as produced being. The point of departure and the result are meant to coincide; for: "In the same way, both the material of labor and the human as subject are the result and the beginning of the movement."[22] Everything appears in the artificial light of technology: "the history of industry and the objective existence [*Dasein*] of industry that has manifested itself [is] the opened book of the *essential forces of the human*."[23]

The positing and posited essence of man rests in that restless "subjectivity of objective essential powers, whose action must therefore also be *objective*. Objective essence manifests itself objectively, and would not manifest itself objectively if objectivity did not lie within its essential nature. It creates and only establishes objects, as it is itself established by objects . . ."[24] However, up until now and due to the undeveloped state of technology, it holds true for the essence of man that "their life expression is also an alienation of life, their realization is an undoing of reality, an *alien* reality."[25] This should cease to be true for the first time during the epoch of thoroughly executed planetary technology. All puzzles, mysteries, and theoretical problems will find their final solution in technical and productive praxis: "All

---

22 Marx, "Nationalökonomie und Philosophie," 237. EN: "Economic and Philosophical Manuscripts," 349.

23 Marx, "Nationalökonomie und Philosophie," 243. EN: "Economic and Philosophical Manuscripts," 354.

24 Marx, "Nationalökonomie und Philosophie," 273. EN: "Economic and Philosophical Manuscripts," 389.

25 Marx, "Nationalökonomie und Philosophie," 239. EN: "Economic and Philosophical Manuscripts," 351.

social life is essentially *practical.* All mysteries caused by theories of mysticism find their rational solution in human praxis and the comprehension of this praxis."[26] Everything is determined by production, by material, sensual, effective, truly objective and concrete praxis: "Religion, family, the state, law, morality, science, art, etc., are merely *particular* modes of production and fall under its general law."[27]

Production is that power which encloses subject and object within the same circle: "Therefore, production produces not only an object for the subject, but also a subject for the object."[28] In this way, subjectivity and objectivity enter into a new relation. The former becomes collectivized and generalized as objective subjectivity; the latter becomes "negated" in its continually renewed status as the product of production; this is not a standing objectivity [*stehende Objektivität*], but instead a becoming *substantiality* [*werdende Gegenständlichkeit*] that constantly reshapes everything that is real.

Planetary technology that has been unleashed and realized in its totality makes literature and art (*techne*) superfluous: "Is Achilles possible with gunpowder and leadshot? Or perhaps the *Illiad* with the printing press or even the printing machine? Does not the speech and song of the Muse necessarily end with the manual printing press . . . ?"[29] The impulse to produce is the driving force; its logical consequence makes everything superfluous that has developed as ideology, superstructure, illusion, romanticism, mysticism, and speculative thought, and become entangled in "so-called world history." Finally, humanity becomes an active

26   Marx, "Thesen über Feuerbach," no. 8. EN: "Concerning Feuerbach," 423.
27   Marx, "Nationalökonomie und Philosophie," 236. EN: "Economic and Philosophical Manuscripts," 349.
28   Marx, "Einleitung zur Kritik der politischen *Ökonomie*" (1857), in *Zur Kritik der politischen Ökonomie*, 247. EN: "Introduction," in *Grundrisse*, 92.
29   Marx, "Einleitung zur Kritik der politischen *Ökonomie*", in *Zur Kritik der politischen Ökonomie*, 269. EN: "Introduction," 111. Axelos uses the German *die Techne* here; not the Greek *tekhne.*

force and exerts real dominance over beings as a whole without
becoming alienated from the products of its labor.

Neither supernatural ideas nor God can oppose the productive activity of man. "Communism makes its start (Owen) with atheism"[30] and leads to the empirical realization of the alienated and metaphysical truth of philosophy. "That which was inner light becomes an all-consuming flame that shifts to the outside. This has the effect that the becoming philosophical of the world is at the same time the becoming worldly of philosophy, that the realization of philosophy is simultaneously its loss, and that which philosophy would battle on the outside is in fact its own internal deficiency . . ."[31] The revolutionary resolution of the alienation of man, the revolution, revolts "against the previous 'production of life' itself, the 'total activity' upon which it was based."[32] Nonetheless, Marx also realizes—although he does not take pause when confronted with this fact—that the radically new cannot remain without hindrances; for: "The supersession of self-estrangement takes the same course as self-estrangement."[33] The revolution describes a circle.

Through this form of thinking, the all of beings is traced back and reduced to the totality of the concrete, sensual, productive praxis of objective human subjects. The world becomes a fabricated world and the issue at hand is: "the sensual world is conceived of as the total living sensual *activity* of the individuals constituting it."[34] The extrasensory (the noetic, the ideal [*das Ideelle, das Ideale*], the transcendental, the metaphysical, the mental) which prevails over (*metá*) the sensual (the real, the empirical, the

30   Marx, "Nationalökonomie und Philosophie," 237. EN: "Economic and Philosophical Manuscripts," 349.
31   Marx, "Aus der Doktordissertation," 17. EN: "Notes to the doctoral Dissertation," 62.
32   Marx, *Die deutsche Ideologie*, 36. EN: *The German Ideology*, 51
33   Marx, "Nationalökonomie und Philosophie," 232. EN: "Economic and Philosophical Manuscripts," 345.
34   Marx, *Die deutsche Ideologie*, 42. EN: *The German Ideology*, 59.

actual, the physical, the material) is overturned and placed on its feet, and as a consequence, everything must be stretched upon the faceplate [of modern technology][35] as a product of sensual (but how can it be meaningful as well?) and productive labor. It is argues here that the realm of quantity, of the number, of the calculating and planning intellect practically generates unheard-of wealth. Metaphysics converts to social physics. Effective "subjectivity" and the effected substantiality, i.e., objective, concretized reality, become the same. The subject is the subject of an object and the object is the object of a subject; "both" are pursued by the same human drives.

What Marx said and thought is actually manifesting itself across the entire surface of the earth-sphere, whether it be in the form of state capitalism or state socialism. Thus, both manifestations move along the same path. We cannot yet know how far these developments will go.

Nietzsche views the same awakening world as a non-world. He conceives of a will to power that encircles the entire mortal coil [*Erdenring*] and desires to make conquest of the universal [*das All*]. The time space of the will *to power* is the game time of nihilism, i.e., of the God who is deceased, as He was killed. The "mad man," i.e., the de-ranged, place-less "planetary" human pronounces the word: "Where has God gone? he called, I will tell all of you! *We killed him* –you and I! All of us are his murderers! But how did we do this? How could we drink up the sea? Who gave us the sponge to wipe away the entire horizon? What did we do as we unchained this earth from its sun? Where is the earth moving to now? Away from all suns? Are we not continually falling? And backwards, sidewards, forwards, to all sides? Do above and beneath still exist? Haven't we gone astray in unending nothing? Doesn't the empty space breathe upon us? Hasn't it grown colder? Doesn't the night and still more night come all the time? Mustn't lanterns be lit in the early morning? Do we still hear none of the

35   TN: Addition for the sake of clarity.

noise of gravediggers who are burying God? Do we still smell none of the divine decay?—for gods decay, too! God is dead! God remains dead! And we killed him."[36]

Will a future mode of being human, will the ascending and descending "*Übermensch*" know how to—and choose—to gain dominance of the entire planet? Will that human entity of a "Caesar with the soul of Christ" be able to *affirm* the eternal recurrence of the selfsame will to power? Will the new humans, who have taken on but also superseded humanity of the past, be correspondingly in bondage to this being (beings as a whole [*Seiendes im Ganzen*]), being that has gone through becoming and is because of it? These questions—should it even be possible to ask them in such a severe manner—can only remain open. The spheres of hermeneutics and the combinatorics of inter-pretations and connections between interpretations seem to gain increasing dominance over the texts themselves. Already Nietzsche *knew* something about this, namely: how every text "disappears" beneath its interpretations.

Heidegger's thought experiment takes up the same questions. "Nietzsche's metaphysics is the completion of philosophy. This means: it has gone through the circle of designated possibilities. A completed metaphysics, which is the foundation for the planetary mode of thought, provides the matrix for an order of the earth that has presumably lasted for a long time. This order no longer requires philosophy, as it is already the foundation of the latter. But the end of philosophy does not also mean the end of thought, for thought is in transition to a different beginning," are state-ments we can read in *Vorträge und Aufsätze*.[37] However, what we must never misconstrue is the way in which the consummation of

---

36   Friedrich Nietzsche, *Die fröhliche Wissenschaft,* no. 125. EN: KSA, vol. 3, 480–81; *The Gay Science,* trans. Walter Kaufmann (New York: Vintage, 1974), 181.

37   Heidegger, "Überwindung der Metaphysik," 83. EN: GA, vol. 7, 81; "Over-coming Metaphysics," 95.

the end, the truth of decline takes place; for: "Decline *lasts* longer than the previous history of metaphysics."[38]

Heidegger attempts to augur a form of thinking that over-comes philosophy—i.e., metaphysics. His preliminary and failing thought questions the distinction prevailing amongst—intra—being as being [*Seins als Sein*] and being as a whole [*Seiendes im Ganzen*]. And this thought unfolds during the age of rising planetary technology and affects man, who has been 'held fast' ['*fest-gestellt*'] as a laboring animal." "Technology" (as com-pleted metaphysics) encompasses "all regions of beings, which respectively equip the whole of beings: objectified nature, culture as an operation, manufactured politics, and over-lying ideals [*die übergebauten Ideale*]. In this context, therefore, 'technology' does not mean isolated spheres of machine equipping and production [*Erzeugung und Zurüstung*]."[39]

Neither an attempt to return to the past, nor remaining captive in the present, neither pessimistic nor optimistic is the open path leading through technology into the future.

III

The planetary does not only determine a particular age, epoch of world history, phase of development, cultural sphere, or stage of civilization. An event is planetary when it allows the being of being as a whole to become established in the history of human essence. Planetary are the course and status of the wandering and erring truth of the world, when within this truth mortals rise, perish, and come again. The level of this event stands under the sign of all that conquers everything, everything drawn into this rotational motion. The plan is that battleground and drafting

---

38   Heidegger, "Überwindung der Metaphysik," 71. EN: GA vol. 7, 69; "Over-coming Metaphysics," 85.

39   Heidegger, "Überwindung der Metaphysik," 80. EN: GA, vol. 7, 78; "Over-coming Metaphysics," 93.

room for that which no plan-maker—whether divine or human—
could realize systematically and without obstruction.

The grounds of the plan for the openness of world-being can by no means be taken into account like the plan of a simple drawing or flat image. Ambiguity and puzzlement are concealed behind the contours of the drawing and the colors of the image. These are the plans that seize and hold humanity, and humans only enter the scene if they have been driven on by planetary blows. Their homeland is uncanny [*Seine Heimat ist unheimlich*], as it is a wandering star, and because this mortal coil [*Erdring*] is not the only entrapment [*Ring*]; it is certainly a ring [*Ring*] that encircles everything, but it remains enclosed within a much more powerful struggle.

The place of the human is the planet earth. Can humanity inhabit this place in a harmonious and planetary way, and find its place and hour within this wandering space of time? The light shone by a non-erring star—can it allow the *universal* [Alles] which is *one* to appear within its radiant clearing and illuminate, ignite, and warm the earthly star, which is not itself a source of light?

Plain and uncomplicated, flat and trite is the way everything appears that exists and is produced in the dawning planetary epoch; planetary actions and productions that plan, level off, and systematize—today and tomorrow—subdue the entire earth: technology becomes the blow that sets everything in motion. By means of technology, the world-historical "essence" of the history of the world as well as the planetary prevalence of the being of beings as a whole in their becoming [*Sein des werdenden Seienden im Ganzen*] unfold. However, this essence appears as a non-essence and pervades in a nihilistic manner. Language, thought, and literature (*logos* and *poiesis*), creative production and empirical action (*poiesis* and *praxis*), art that has become technology (*technê*), and labor in general as the production of *erga* [works], no longer provide any evident meaning. Truth, the meaning of being, remains absent. The extrasensory has been degraded, the sensual has been stretched upon the faceplate, and everything has become a product of monstrous fabrication; by no means, however, will rational and artificial purposefulness and goal-orientation be able to truthfully overcome the meaningless and the aimless. Such planning and production, this activity and this business, these compulsions and these total operations, this mobilization, mechanization, and intrigues (of power), determine the continuing and progressive use of all produced and useful objects—of beings as a whole. The expedient and useful are consumed and exploited in an increasingly rapid tempo. This is in order and is part of the persevering planetary order.

Zeus guides all that exists with lightning bolts. The illuminating and destructive fate of the bolt does not open the horizon of the *one-all* anymore. "Beings as a whole are guided by the bolt," Heraclitus says,[40] speaking of the divine lightning bolt. Zeus has become Jupiter, and Jupiter has become a planet. Driven and beaten, but not at all guided, mortals can no longer say about divine being, about the *holon*, about that which is sacred: "The one, that which alone is wise, cannot be named and certainly cannot be given the name of Zeus";[41] they themselves strive to be bearers of the planetary will to power, of the will to will.

Ares no longer leads sacred battles [*Kämpfe*]. The Heraclitean *polemos* ("the father of the whole of beings") has become the planetary state of war and only has the naked domination of the earth as its aimless objective.[42] Beneath the planetary course of the stars, the struggle for the whole of the visible earth and its measured spaces of time remain entrapped in void nothing. Peace is just as profane as war, and perhaps the will to power has also lost its belligerent courage and can only play out in a belligerent state of peace.

---

40   Heraclitus, frag. 64. EN: "Heraclitus," 104.

41   Heraclitus, frag. 32. EN: "Heraclitus," 119.

42   EN: This is a reference to Heraclitus, frag. 53; "Heraclitus," 102.

Aphrodite—as Venus—has become a planet. Eros has turned into love, and love has become a matter of the imagination and of action. Love is replenished by the powers of self-reflection, executed as an act and as reproduction. Marx already conceived of "being together [*Mitsein*]," Greek: *synousía*, in a "technical" mode; he perceived it from the vantage point of the division of labor; the "division of labor, which was originally nothing more than the division of labor during the sexual act."[43] Future technological production of "human" creatures is not far away; the reproduction of the human race will be reduced to the parameters of total production. The planned and artificial breeding of "rational living beings" would merely be another of the many steps of *physis* transitioning into *technê*.

But after the entire void has been filled according to plan, after the entire surface of the earth's crust (and the underlying layer of mother earth) as well as the air and the seas surrounding it have been processed with technological means, after the elementary forces—both the smallest of the small and the greatest of the great—have been released, i.e., "mastered"—what happens then? After the entire sense of these "nonsensical" goings-on have been realized at the planetary level—in the north as in the south, in the west as in the east—will meaning collapse after the decline, or even shipwreck, of the future world?

After the abolition of any meaning, and after the fabricated planetary world has become constructed in its techno-aesthetic mode, will the openness of meaning, the wandering and erring truth of the being of beings as a whole (provided all of that will even be described in this way and require these designations) reveal itself? After planetary humanity has been tormented to the point of exhaustion by nihilistic, meaningless, all-round business and the aimless practical compulsion to produce, will nostalgia and the yearning for faraway places be able to coincide with one another?

43   Marx, *Die deutsche Ideologie*, 28. EN: *The German Ideology*, 42–43.

The *logos* of Heraclitus, whose traces lead to planetary thought, once conceived of the world of being of worldly time as a "game." He is the first—Occidental—thinker who dared to do this. Two and a half centuries later, during the late period of an epoch drawing to a close, and during the early phase of a newly dawning era, Nietzsche calls attention to the higher meaning of a world game that is neither meaningful nor meaningless. Heraclitus and Nietzsche are the only thinkers who dared to place the being of becoming on the gameboard [*Spielbrett*]. Perhaps the day after tomorrow, planetary nihilism that has been overcome will be able to hear its own voice anew. Indeed, Heraclitus said: "Worldly time is a child at play—a child with a board game [*Brettspiel*]: the kingdom of a child."[44]

And Nietzsche permits the "mad" man to ask, after he has announced the murder of God: "How can we console ourselves, the murderers of all murderers? The most sacred and the most powerful the world has ever possessed has bled to death beneath our knives—who can wipe the blood from ourselves? What water could we cleanse ourselves with? What ceremonies of atonement, what holy games will we have to invent?"[45] This game of the sacred and the profane—could it be the "world" itself, a world of gods and man overcome? "Around the hero everything can become tragedy, around a demi-god everything can become a satyr play; and around God everything becomes– what? Perhaps a 'world'?"[46]

After the all of beings, the totality, the being of the becoming world has "lost" the meaning of its openness, after the senseless-ness of planetary technology has fully manifested and exhausted itself both in terms of meaningfulness and method, perhaps the

44   Heraclitus, frag. 52. EN: "Heraclitus," 102.
45   Nietzsche, *Die fröhliche Wissenschaft*, no. 125. EN: KSA, vol. 3, 481; *The Gay Science*, 181.
46   Friedrich Nietzsche, *Jenseits von Gut und Böse*, no. 150. EN: KSA, vol. 5, 99; *Beyond Good and Evil*, in *Beyond Good and Evil/On the Genealogy of Morality*, trans. Adrian Del Caro (Stanford: Stanford University Press, 2014), 52.

one-all itself (the *hen pánta*), the holon as the holy, inspirational essence, and the truth of being as game will be able to unfold their harmoniously martial powers. Perhaps a holy-*profane* game can bring into play the "non-being" of the openness of the being of beings as a whole, the horizon of the rotational motion of the world, the difference itself of that which is distinct from being in genesis, the never concluded and never completed totality, and even the course of the planet—a deadly serious world-game into which mortals are playfully plunged in play. In this way, the harmonious and planetary essence of world-being would appear in the playroom of "time without aim" neither as a tragedy nor as a comedy, but as an "open world."

Twelve Fragmentary Propositions Concerning the Issue of Revolutionary Praxis

I

Marxism—and the "Marxisms" that have been increasingly combined with other elements: Christian, bourgeois, positivist, scientistic, psychoanalytic, phenomenological, existentialist, structuralist—can still serve as incentives for certain types of theoretical research without, however, providing the decisive impulse for revolutionary praxis in highly industrialized societies, as this is intended in accordance with Marxian and Marxist schemata. Marxism and Marxisms execute moderately productive and inquisitive labor, they become integrated in the theory and praxis of their society, and they live out its life and death.

II

Highly industrialized society is gradually being transformed into a "socio-capitalistic" society, set in motion by an ever-advancing homogenizing technology.

III

In technologically undeveloped countries where Marxism has not avoided confusion by merging with different elements—religious, ethnic, nationalistic, ideological—, it has still retained the role of a lever for certain revolutionary changes which lead those countries to a socio-capitalistic and techno-bureaucratic status quo, to a planetary society according to plan.

IV

Sociological studies that call themselves Marxist, or intend to be Marxist, are increasingly losing their autonomy [*Eigenständigkeit*]; they become confounded with other types of research and the result is oversimplification. The theoretical whole of Marxism—in its specificity and as a totality—has ceased to exist, whereas theoretical Marxism is undergoing negation in a weakened and non-offensive form through its own self-generalization. It has been assimilated by the wave of amoebic progressivism [*von der Welle des teigigen Progressismus*].

V

Marxist analysis has retreated from its own potential—to an equal extent in capitalist countries and countries proud of their socialism. Marxism does not place itself in question sufficiently enough and does not ask radical questions.

VI

The famous "changing the world" is taking place nonetheless, though in accordance with a mixed and impure schema. The designs of innovators, of reformers and "revolutionaries" have themselves become transformed in and through historical momentum, which can only proceed with misunderstandings and compromises.

VII

Socialist movements have failed, as every radical intention "fails," because it must manifest itself through approximations. Their conception itself was too ideological and too abstract; it did not think profoundly and misjudged its origins, its proper procedure, and its aim. As it was too utopian, this conception was con- demned to fail the test of banality; as it was too banal itself, it was unable to uphold its utopia as an infantile and eschatological hope. From the onset, the revolutionary and socialist conception remained determined by that which it claimed to negate.

VIII

The so-called socialist bloc—which is already polycentric to a significant degree—seems to insult those who want to be open Marxists and socialists, democratic and liberal communists. They are incapable of comprehending the play of social realities, the role played by violence, by oppression, and by the state. They overestimate and underestimate what is said and done without rightly knowing how to interpret it.

IX

Highly industrialized societies are heading for a capitalistic socialism of the state [*einen kapitalistischen Sozialismus des Staates*]; they preserve their rule, power, and exploitation in ways that have become increasingly more mediated, and they let us see with grim foreboding that the complete self-admin- istration of socialized society—the negation of dominance—is a myth. Through its processes of collectivization and univer- salization, bourgeois society socializes individualism, engulfs and neutralizes all attacks, barely succeeds in integrating every criticism, emasculates and acknowledges uncompleted revolts by dissolving the distinction between true and false, freedom and unfreedom.

X

The left is merely able to participate in the historical comedy of bureaucratic nationalization; its own tasks have become intermingled with general duties that simultaneously express individual and collective interests. When the left adopts a negative position, it achieves nothing; if the adopted position is too positive, the left is incorporated by the establishment. If the left pursues a mainstream course, it contributes to the prevailing mediocrity. Aggravating social antagonisms and contradictions does not appear to be the lot of the left, as the overall status quo is able to digest such contradictions and antagonisms with their originators all at the same time. The protests of the left remain ineffective and empty; they enthusiastically maintain the voice of a certain aspiration, but this aspiration remains imprecise.

XI

The proletariat in technologically developed countries has become assimilated into the general scheme of things as petty bourgeoisie; its being and consciousness negate themselves through the process which leads towards general prosperity. The engine of this process—whether capitalistic or socialistic— remains profit. Complex crises cannot be predicted, and the revolutionary praxis of "advanced" countries cannot be executed by theoreticians and professionals of a revolution that never arrives. Revolution does not even seem to be a remote possibility and the class conflict—which has become extraordinarily sluggish—does not have revolution as its horizon any longer. The practices of reform and modernization take place without magnificence. The "underdeveloped" countries will soon attain the achievements of the bourgeois French Revolution; Marxism is the instrument of their emancipation and industrialization.

XII

Marxian and Marxist theory and their practical perspectives remain caught between the anvil of Hegelian political philosophy and that which it expresses, and the hammer of the nihilistic diagnosis formulated by Nietzsche (liberated from appeals to the spirit and all romanticism). While the planetary age continues its odyssey, attempts to gain expression, and changes its actors. Negativity conceals itself during this time with no difficulty.

A Discussion about Science[1]

Conducted with the Classical Philologist Jean Bollack

Jean Bollack: Conspicuous these days is a shared concern about the threat posed by science, connected with the expansion and perfection of technological methods. Investigating the material occurs within parameters that exceed the powers of an individual researcher and by means of technical equipment, statistical analysis, and the assessment of quantitative units. It is indeed the case that nobody can avoid this positive enrichment within their own field of knowledge, and in spite of this everyone has the feeling that their area of knowledge evades overall accessibility, and as a result, evades their own purview; and this explains another, not less conspicuous effort, namely, to rediscover the unity of science or even several distinct disciplines and give them a common goal. Didn't the geographer speak of merging geography with history and sociology, and Lévi-Strauss, whose most important achievement—the description of kinship systems—involved

1 EN: The German *Wissenschaft*, which we have here translated as "science," has a somewhat broader meaning than the English word, and can mean scholarship in general.

the application of structural linguistics to ethnology, didn't he believe he had foreseen a general functional theory of human cognition?

Kostas Axelos: Yes. But at the same time, scientific research has become a kind of executed technique.[2]

JB: I don't understand your "but." Technology does seem to me to be one of the two main characteristics that I mentioned.

KA: Technology is still understood in a sense that is too restricted. The sciences cannot take complete possession of technology, as technology constitutes their driving force, the power that guides them. In general, one regards technology to be the product of science. But isn't it much rather the most internal impetus [*Beweggrund*] of science? Not only is the research of science carried out in a technical manner, but the field itself and the objective of every science are determined by—scientific—technology. Ethnology studies its object exhaustively with technical means: so-called primitive peoples are given expression through the technical language of ethnology.

JB: That's quite right. In the view of historians, all of world history has become the history of economics. Products and trade also dominate the feudal era now—one only needs to think of the frequently stressed significance of the studies by Marc Bloch about feudal society—, yes, even antiquity, while on the other hand obscure and shadowy individuals have been promoted to the status of movers of history, as they are resurrected like phantoms from historical statistics and archive research—pale and unassuming, a strange Last Judgment, no monarchs, no lords, no popes and cardinals,

2   TN: See Bruzina, "Translator's Introduction," in Axelos, *Alienation, Praxis and Techne*, ix–xxxiii for a discussion of the special use made of "technique" by Axelos, as opposed to "technology."

but a countless horde of "rural population" and "urban
bourgeoisie," market and anonymity.

KA: What isn't offered on the market these days? Even the unusual has become a commodity. Time is reflected by history itself.

JB: To that extent, research had been affected by the spirit of the times, before it acquired a methodology. I also admit that the desire for inclusion itself still lies within the sphere of technology. Of course, Lévi-Strauss speaks of a theory of *fonctionnement* of the human spirit: machine and automation are still mentioned in his last theory. And the *sciences humaines*, as one calls organized research in the humanities in France nowadays, are certainly always sciences *of* humanity and never *for* humanity [der *Menschen und nie für* den *Menschen*]. How does the human behave, how does their mind function?—these are the questions that are asked. And in spite of this, in my opinion, there lives in France—as was indicated by my summary just now—the idea of, or at least the desire for research that is still humane, accessible as a whole, and goal-oriented at the same time. Questions concerning object and method are consistently addressed. It seems to me, it is precisely in the determination of this goal that the whole problem is rooted. As this goal is only still contemplative to a marginal degree (in the tradition of the liberal arts), does it not necessarily intrude into the activities, yes into the destiny of a nation? In this regard, it is characteristic that many commentators view science as action, as an activity that advances, whereas research is considered to be the supplementary execution of the overall occurrence. This is especially striking in the field of scientific research where reactionary interests can simultaneously constrain scientific investigation and economic recovery.

KA: They advance at the same pace as contemporary science.

JB: Not at all. I would even maintain—always the same.

KA: But here, the same bears different names. You frequently spoke of research. That seems to me to be a very important point. Today, science has become research. There aren't any more well-rounded, organically promoted and structured individual scientific disciplines. Classical physics does not exist. All of us know that. And it's already begun to be the case that psychology, sociology, and a uniform conception of history are ceasing to exist. Research is the predominant reality; constantly renewed research that desires to know less than to investigate and produce effects. For this reason, all scientific institutes are called research institutes; for the same reason, the image of the scientist and scholar has changed radically. Even philosophers are called researchers and paid as such by the state. In research activity, however, lies the power of a technology that constantly presses forward. The human as subject, and material objects as the focus of research, exceed the subject-object dichotomy which holds for the classical sciences, and they enter a new, apparently simple albeit four-dimensional zone.

JB: In my view, all programs seem to confirm how much science has lost its national character. Statistics remain statistics, whether they apply to the Roman Empire or to Bantu tribes, whether they are calculated in Kiev or in Paris. But regardless of this, I did call attention to tendencies typical for France just a while ago—to conceive of science as human and utilitarian in the best sense of the word, which is not identical with its absolute effect, and perhaps this reflection is the best antidote for an anonymous effectiveness that evades human intervention. What, however, we can continue to evaluate quite positively—in my opinion—is the outstanding role that can still be attributed to individual persons. Don't the same names reoccur in different contexts? Or are they merely the avant-garde of a radical efficiency?

KA: The technological and scientific research spreading across the entire earth-globe [*Erdball*] has managed to retain its own character in every country. French researchers still consider themselves to be humanists. But perhaps modern *planetary* research—including the French—is neither natural nor unnatural, neither human nor inhumane, neither purposeful nor without purpose. The magnificent aspect of contemporary science is precisely the resolution of these oppositions. The whole affair can appear meaningless, but it isn't. But it is not all meaningful, either. It "is," as it constantly dissolves, becomes fixated, and negates, and thereby sets the planet earth in motion. The desire you refer to, which is actually more a desire for the humane, seems to be a desire for the poetic in my opinion. While pursuing such a problematical anthropology, isn't it precisely literature that Foucault encounters?

JB: You rightfully emphasize the diction of the text, and at the same time, you identify the problem of language. Most submissions stand out due to the clarity of their composition and their expressive finesse, which still remain virtues of the French intellect. All the same, one notices a certain perplexity that they possess. The academic jargon, the form of expression of scholarly circles, sounds outmoded, and conspicuously often one has the impression that material has been borrowed from technical functions, and "*l'en fonction de*" expresses the reason—as if the ascertainment should receive more functionalistic than merely positivistic resonance. Is this a stage on the way to encoded statements with restricted communicativity?

KA: In a curious way, the functionalistic application of language has connections to the Romanic *clarté*. Many times a statement seems to be so clear that one cannot understand it in a productive manner any more. I must state quite openly: when one speaks of sociology today, I don't know whether one means the nearly antiquated sociology of the nineteenth century, our contemporary dilemma that is so confused, or the orientation of research intended to surpass traditional school sociology. The clarity and sobriety of exposition also frequently prevents the surfacing of more profound levels of reality and developments of thought. The so-called philosophical reflection on a theme is not necessarily tied to a form of investigative and creative thought.

JB: Well, this broadcasting series was concerned with research reports and only secondarily with autonomous analysis.

KA: But shouldn't the report itself have a research character? Why do so many researchers nowadays fear the accusation that something is unscientific?

JB: I would say: one no longer knows the limits of scientific discovery with any precision, and one becomes even more constrained than one would actually care to be, or one seeks refuge in the pure correspondence of technical procedure, in encoded statements, as this has already begun to take root in the humanities as well.

KA: All newly discovered knowledge—think of Marx, think of Freud—was branded as arbitrary by academic minds. Perhaps there are no borders between science and inquisitive research, if both are authentic. Borders can only be drawn between dogmatized and open knowledge.

JB: Agreed. But is the scientific professionalism mentioned before—as the criterion for academic performance— identical with what you mean by "dogmatized science"?

KA: The greatest part of academic performance takes place within a previously determined space. In one way or other, academic research is coerced from the outside to open its horizons.

JB: Is an open Scholasticism in your sense of the term even con- ceivable? Mustn't the inherited norm—as it is still handed down in France, for example—coexist with the collapse? It is characteristic that the most intellectual influences of European rank in Paris—and this since the eighteenth century—have come about outside of the traditional schools of thought. Bergson was never at the Sorbonne and Sartre never held an academic chair. Even students lead a double life between the burden of university duties (with countless examinations) and their second free and more or less "present" existence. Moreover, this also explains the education of French intellectuals that often appears so divided from without (but characteristically not in respect to academicians; the difference is significant); they often represent tradition and revolutionary spirit at the same time. In cases where this discrepancy cannot be tolerated any more, this seems to be the reason for the attraction that orthodox Marxism is still able to exert, in my opinion. All of this exists in very stark contrast to Germany where the university still focuses intellectual life, even if it does not fulfill it.

KA: An open Scholasticism doesn't exist. However, as soon as Scholasticism—or all current forms of neo-Scholasticism (of all variations)—are experienced as such, they serve the function of inciting their own forceful revocation. The renowned physicist Louis de Broglie once related how his teacher told him after the first phase of his school education that he had no talent for physics research. And this duel (tradition versus violent overcoming) produced modern physics. But a new danger exists: that achievements become degraded once again to generic school material. For example, in all universities today one studies as an object of research—without properly reconstruing the matter—the literary thought of André Breton, who is not sufficiently acknowledged as the founder of Surrealism, or—in Germany—Hölderlin, who has become a figure of the all too academic history of literature and literary science.

JB: That is an old and general fate. The thought of Heraclitus became Heraclitism and Plato's became Platonism. A fate that was taken *ad absurdum* only by organized technical operation and generalized curiosity. No community has ever been able to avoid epigonic dogmatism. But today, normative dogmas that are still alive no longer exist at universities, which is perhaps something new and something that can be evaluated positively. Perhaps we have exited the old repetitive cycle of creation and codification, and the open quality you speak of is probably not any radically new foundation any more like the architectonic of medieval Scholasticism by Thomas Aquinas, or the methodology of Descartes in Cartesianism. Broglie, whom you mentioned, ventured into the uncertain. He cannot be the founder of a school of thought, as heroic as his attempt at systematization might appear. And not Einstein, either.

KA: Certainly, the researcher of today swims in a different stream. It has been emphasized enough that one cannot and should not separate the physiological experiment from the organic conductor of the experiment. Here, the subject-object dichotomy is in the way to its "overcoming." The human subject and objects [*Gegenstände*] at the focus of research (so-called objects [*Objekte*]) become negated at the same time and make room for the appearance of new content. Now, I repeat the question: In what relation do these efforts stand in respect to poetry? Do dry science and fiery poetry remain separate, even irreconcilable?

JB: In this way, we come back to the theme of programs with a poetic presentation.

KA: Still more important is evocative literature itself.

JB: Yes, Foucault was a proponent of the fulminatory, winged Heraclitean poetry of René Char, and yet as an anthropologist he treated Teilhard's protohistoric hypotheses and Lévi-Strauss's structures of kinship systems as communicative forms. The literary expression is understood as conveying the meaning of the theory, so to speak. "From the onset," as he phrases it so elegantly,

"man is doomed to meaning," man exists in "that uncanny
hyperbole of constantly replicated signs that constantly
function as symbols in relation to one another, of such signs
that mutually observe each other furtively and reply to
one another in a language with the same vocabulary as the
silence of the nights."[3]

KA: How strange that statement sounds about the voice of
silence.

JB: Everything remains unexpressed and simultaneously
becomes communication and correspondence. Many seek
a determination of values, the meaning of denotation,
signification as meaning and denotation in sociology. Others
even go a step further; language becomes a reflection that
ascertains meaning for them. They alone are capable of
discovering specific relations. The definition of man as *homo
loquens* acquires a different sense: humanity stands before
a matrix of symbolic relations. Just as a code remains silent
within the speechlessness of its technical transcription, literary diction liberates ciphers that bear their own decoding
within themselves. The bifurcation between logical thought
and imagination dissolves, indeed, into a logic of multiple
levels. The difference between primary and secondary reality
disappears on the spot. In this way, the present (or future?)
language intended to investigate the original propositions
of the pre-Socratics returns to where neither the dark,
perplexity, and light of Heraclitus, nor the eulogy and the
directive of Empedocles were distinct from one another. And
even with the Plato who founded the Academy, one finds
myth and logos that supplement one another. In another
such fundamental correspondence, André Breton, the
unerring sustainer [*Wahrer*] of the emancipatory, who you
mentioned previously, put—as Freud's successor—magical
culture and the actuality of derivative realities in their

3   EN: As far as I am aware, Foucault's contribution has never been published.

legitimate place. In his—as one phrases it—materialistic theories of "objective accidents" lies the geometrical locus of determinative coincidences, a locus, as he himself stresses, that is identical with Hegel's sovereign methodology.

KA: That is important. Naturally, Breton interprets Hegelian dialectic in an open manner.

JB: Liberated powers of imagination become "scientific" as an outflux of analyzable faculties within the realm of dreams and unrestricted mind. Human necessity becomes natural necessity, and thus, science and literature meet within "objective chance."

KA: How and where is the decision made today? What is chance and what is fate? The duel between thought and logic, soul and psychology, historical humanity and sociology has not been initiated in the least. Man as a sign with which the world plays appears as a number in the age of technology. Research based on calculation and functionalistic knowledge play with ever greater numbers. Literature and thought await the hour of their encounter. Already Hegel drew the distinction between reasoning thinking and speculative thinking. In philosophy seminars, this distinction itself is investigated with the faculty of reasoning thought. Up until now, we have spoken of researchers and scientists. But let us also allow a graphic artist to take the podium.

JB: I anticipate already: you want to draw our attention to the remarks of the painter Braque that had ripened under the brilliant light of graphic observation, as he stated that thought and reasoning were two different things: *Penser et raisonner font deux* [thinking and reason are two things]. For a painter, graphic perspective and poetic sense constitute thinking, too.

KA: It is no accident that the wide-ranging perspective within an increasingly structured, functionalized, and automated interpretation of the world will attempt to see through its own "free" zone. Structuring, functionalization, automation are the driving planetary slogans in the West as well as in the East, in the North as well as in the South. The thought corresponding to this world situation can, however, transcend all particular dimensions by experiencing the unity in diversity and giving expression to it. This new strictness of thought will probably—of necessity—supersede every codified or even codifiable logic.

JB: Accordingly, the technicized and even automated cipher would achieve the same things as the poetically receptive diction mentioned beforehand. Just as no word or collection of words can stand for itself any more in a linguistics that has become structural, and just as the class of their denotations only achieves meaning through the play of relations among them, the system Lévi-Strauss developed for extremely confined societies also makes reference to an unsuspected, incomprehensible, and—apparently only manifested as— mechanical interplay of more comprehensive groups. In this way, its already complicated schemata become constellations on the illuminated screen of blinking points within the sphere of the inestimable (even if this visibility is made possible by a machine) and structure becomes automation. And as a result, a new light is thrown on the remark made at the beginning that ethnology designs our own hardship.

KA: The thought of the future, which has already somehow managed to break through in our present, will also dare to place all scientific achievements in question. Language is by no means coextensive with linguistics and there are only questioning answers to the great questions. Are we moving into the realm of that "silence of the nights"? Or is the light dawning–and the language emerging—of new heavenly motions?

Epilogue

The thought experiment [*Denkversuch*] attempted here dares to seek orientation in the world game in order to give expression to the game "itself" and play it. In its uniformity and reference to the whole, such an attempt can only unfold in a fragmentary and spotty manner. Various modes of speech must be tested: scientific, poetic and intellectual—systematic and aphoristic—, historical and historiographical [*geschichtliche und historische*]. All of these directives can only be carried out with imperfection. Whether we want this or like this—or not. With and without success. It is and remains our task to enter the planetary age. Striding forwards and backwards. "Thoughts that come on the feet of doves guide the world," were once the words of Nietzsche.[1] Do we already sense that such thoughts that guide the world—and still more: thoughts guided by the world—stand before our doors?

Mythical—or mythological—prehistory, the primitive and the archaic confront us with puzzles. The Eastern [*Das Morgenländische*], the Oriental, and the Asian stand behind us, still questionable, terrible, and fruitful, behind "our" Occident. The Greeks and the reconstructive-constructive Romans still stand before us—they question us and are questioned by us. Judaeo-Christian values are spreading and becoming extinct at the same time, as the last and final, perfected religion taken over by secularism, humanism, and socialism. The European and modern, modernity, is still an active principle, universalizing itself and remaining a destiny. The planetary draws us into its storm.

As humanity has been drawn into storms from time immemorial, we make the attempt to order everything and construe logical and ontological as well as ontic schemata and realms of

---

1   EN: Friedrich Nietzsche, "Also sprach Zarathustra II," in KSA, vol. 4, 189; "Thus Spoke Zarathustra: Second Part," ed. and trans. Walter Kaufmann, in *The Portable Nietzsche* (New York: Penguin, 1954), 258.

experience. The logical not only constitutes a propaedeutics of speech and thinking. In its guise, the onto-theo-logical grows and fades, i.e., the philosophical and metaphysical: it desires to experience being. But being always appears as beings—*God, universe, human—,* and as a result the theological, the cosmological, and the anthropological allow questions about the divine, the natural, and the human to emerge and expire. The ethical, the political, the poetic and artistic still manage to accomplish this, whereas nearly omnipresent technical developments require technology. Every circle encircles the others and is encircled by them. God (*divine logos*), nature (*the cosmic order*), man (*the beginning, middle, or endpoint*) revolve in circles. Each of these three powers can be the beginning and be identifiable at the end. *Divine logos* (and a "logical" God), *cosmic nature*, as well as thinking and acting man are the only three thoughts which seem to be at the disposal of humanity in order to experience being and Dasein, or everything and nothing in either a hard or approximate manner. Historical epochs as well as diverse theoretical and practical dimensions have worked them out. Nevertheless, they remain unthought-of. *Each* of these "three" powers appears as the secret that evolves into the other. Even when one places them together, encloses them in a revolving circle, or comprehends them as the circle itself, they fail. Beginning and end blend without clarity. The Greek and Judaeo-Christian, Hegelian, evolutionary, or Marxist three-step cannot rid themselves of their limp. Does it begin with divine or dialectical logos or a "logical" God, and proceed from there to a history of humanity that generates, investigates, and recognizes everything? And so forth . . . The implicitly or explicitly dominant pattern of thought here is too trivial.

Within and as a world history that encompasses cosmic nature and the history of humanity in one ring and one struggle [*Ring und Ringen*]—although each of these (two?) powers encloses the other—, belief and action, speech and thinking, work, love, battle and death *unfold*, or interpreted in one problematical word: *play*

[*Spiel*].[2] Myths and religion, literature and art, politics, philosophy, science and technology—technicized natural sciences and humanities—are their reifications and institutions. Proceeding from the family, to other concentric and well-defined institutions, to eccentric and combinatory games, up to the future world state, everything called "life" manifests itself. We have not yet experienced the gravity and void of institutions, nor fathomed their past, present, and future.

What—or who—will be brought into the game of motion—of time, of errance? Who, or what, or which ones will be put at risk within the space of game and time known as humanity and the world? In respect to which origins and which future? A matrix of convergent questions, as well as productive and receptive, repetitive and problematical conquests, compel us towards "new" thought experiments and experiences of the world. Urgent questions are involved here, even in contexts where the answers have become untenable. This about a more radical understanding of truth, and about constellations that triumph as reality and then perish, and which do *correspond* to errance.

This is about overcoming metaphysics, the supersession of that foundation, and of *that* meaning—and at the same time: of the nihilism—which negates being, foundations, meaning, *and it itself*. This is about overcoming subjectivity and objectivity. This is about the end of history and traditional humanity. These questions can contribute to the openness of the game and to new closures.

But whose affair is this mode of thought nowadays, and how are the world game and constituted worlds experienced? Which individuals, which nations, and which world history do not avoid this challenge? Which stars perish, which constellations ascend and shine?

---

2   TN: Italics added by the translator for the sake of clarity.

Two modern summits of the European Occidental world, France and Germany, do they think and are they alive? Thinking—if such a thing does indeed exist—has not been a French problem since Descartes and Pascal. Germany, in its nostalgia for Greece, has taken it up. But life—or that which one considers to be life, so perplexing and gigantic—does not appear to be a German reality. After he has abandoned Greece with disappointment and arrived amongst the Germans, *Hyperion* writes to Bellarmin: "I cannot imagine a people more torn apart than the Germans. You see craftsmen, but no humans, priests, but no humans, masters and slaves, young and mature, but no humans—isn't it like a battlefield where hands and arms and limbs lie about in pieces all on top of one another, whilst the shed blood of life sinks into the sand?" Is the game denied to Germany? But which game? "Your Germans like to halt at the most essential, and for that reason there is also so much bungling effort among them, and so little of the free and truly joyful."[3]

Above France and Germany stands Europe, neither being nor not being, as well as its two branches, America and Russia, drawn into a common planetary melting pot. The same process is also occurring in all other parts of the world. Does this mean: world history above everything? Does the *world game* "stand" in it and "above" it, and is it open for the planetary thought of tomorrow? The game pervades all of world history, unsettling and igniting, dissolving and smashing. It bears and then squanders a few names and ventures. Does the world assemble everything in its playful mode? Everything appears and breaks apart within the game of world and humanity. "Is" the game only one of the puzzles of the world, or "is" the world only a form of the game?

---

3   EN: Hölderlin, "Hyperion oder Der Eremit in Griechenland," 168; "Hyperion, or the Hermit in Greece," 128.

Works Cited

Axelos, Kostas. *Héraclite et la philosophie: la première saisie de l'être en devenir de la totalité.* Paris: Éditions de Minuit, 1962.

———. *Marx penseur de la technique: De l'aliénation de l'homme à la conquête du monde.* 2 vols. 1961. Paris: Éditions de Minuit, 1974.

    Alienation, Praxis and Techne in the Thought of Karl Marx. Translated by Ronald Bruzina. Austin: University of Texas Press, 1976.

———. *Vers la pensée planétaire: le devenir-pensée du monde et le devenir-monde de la pensée.* Paris: Éditions de Minuit, 1964

Heidegger, Martin. *Aus der Erfahrung des Denkens.* Pfullingen: G. Neske, 1954.

    "Aus der Erfahrung des Denkens." In GA. Vol. 13, *Aus der Erfahrung des Denkens 1910–1976,* 75–86.

    "The Thinker as Poet." In *Poetry, Language, Thought,* translated and intruduced by Albert Hofstadter, 1–14. New York: Harper Perennial, 2001.

———. "Bauen Wohnen Denken." In *Vorträge und Aufsätze,* 145–62. Pfullingen: G. Neske, 1954.

    "Bauen Wohnen Denken." In GA. Vol. 7, *Vorträge und Aufsätze,* 145–64.

    "Building Dwelling Thinking." In *Poetry, Language, Thought,* translated and introduced by Albert Hofstadter, 145–61. New York: Harper Perennial, 2001.

———. "Das Ding." In *Vorträge und Aufsätze,* 163–81.

    "Das Ding." In GA. Vol. 7, 165–87.

    "The Thing." In *Poetry, Language, Thought,* translated and introduced by Albert Hofstadter, 161–84. New York: Harper Perennial, 2001.

———. *Einführung in die Metaphysik,* Tübingen: Niemeyer, 1953.

    GA. Vol. 40, *Einführung in die Metaphysik.*

    Introduction to Metaphysics. Translated by Gregory Fried and Richard Polt. New Haven: Yale University Press, 2000.

———. *Der Feldweg.* Frankfurt am Main: Vittorio Klausmann, 1953.

    "Der Feldweg." In GA. Vol. 13, 87–90.

    "The Pathway." Translated by Thomas F. O'Meara. Revised by Thomas Sheehan. In *Heidegger: The Man and the Thinker,* edited by Thomas Sheehan, 69–72. New Brunswick, NY: Transaction Publishers, 1981.

———. "Die Frage nach der Technik." In *Vorträge und Aufsätze,* 9–40. Pfullingen: G. Neske, 1954.

    "Die Frage nach der Technik." In GA. Vol. 7, 5–36.

    "The Question Concerning Technology." In *The Question Concerning Technology and Other Essays,* translated by William Lovitt, 3–35. New York: Harper & Row, 1977.

———. "Grundsätze des Denkens." *Jahrbuch für Psychologie und Psychotherapie* 6, no. 1/3 (1958): 33–41.

    "Grundsätze des Denkens." In GA. Vol. 79, *Bremer und Freiburger Vorträge: 1. Einblick in das was ist 2. Grundsätze des Denkens,* 81–175.

170

"Grundsätze des Denkens." In. GA. Vol 11: *Identität und Differenz*, 125–40.

"Basic Principles of Thinking." In *Bremen and Freiburg Lectures: Insight Into That Which Is and Basic Principles of Thinking*, translated by Andrew J. Mitchell. Bloomington, IN: Indiana University Press, 2012.

———. "Hegels Begriff der Erfahrung." In *Holzwege*, 105–92. Frankfurt am Main: Vittorio Klostermann, 1950.

"Hegels Begriff der Erfahrung." In GA. Vol. 5, 115–208.

"Hegel's Concept of Experience." In *Off the Beaten Track*, edited and translated by Julian Young and Kenneth Haynes, 86–156. Cambridge: Cambridge University Press.

———. "Identität und Differenz." In: GA. Vol. 11, *Identität und Differenz*, 27–110.

Identity and Difference. Translated by Joan Stambaugh. 1969. Chicago: University of Chicago Press, 2002.

———. *Nietzsche*. Vol. 2. Pfullingen: G. Neske, 1961.

GA. Vol. 6.2: *Nietzsche II*.

Nietzsche. Vol. 4. *Nihilism*. Edited by David Farrel Krell. Translated by Frank A. Capuzzi. San Francisco: Harper & Row, 1982.

———. "Nietzsches Wort 'Gott ist tot.'" In *Holzwege*, 193–247. Frankfurt am Main: Vittorio Klostermann, 1950.

"Nietzsches Wort 'Gott ist tot.'" In GA Vol. 5, 209–267. "Nietzsche's Word 'God is Dead'." In *Off the Beaten Track*, edited and translated by Julian Young and Kenneth Haynes, 1–56. Cambridge: Cambridge University Press.

———. *Der Satz vom Grund*. Pfullingen: G. Neske, 1957.

GA. Vol. 10, *Der Satz vom Grund*.

The Principle of Reason. Translated by Reginald Lilly. Bloomington: Indiana University Press, 1996.

———. *Sein und Zeit*. Halle: M. Niemeyer, 1927.

———. "Der Spruch des Anaximander." In *Holzwege*, 296–343. Frankfurt am Main: Vittorio Klostermann 1950.

"Der Spruch des Anaximander." In GA. Vol. 5, *Holzwege*, 321–73.

"Anaximander's Saying." In *Off the Beaten Track*, 241–81.

———. *Unterwegs zur Sprache*. Pfullingen: G. Neske, 1959.

GA. Vol. 12, *Unterwegs zur Sprache*.

On the Way to Language. Translated by Peter D. Hertz. San Francisco: Harper, 1971.

———. "Der Ursprung des Kunstwerkes." In *Holzwege*, 7–68. Frankfurt am Main: Vittorio Klostermann, 1950.

"Der Ursprung des Kunstwerks." In GA. Vol. 5, 1–74.

"The Origin of the Work of Art." In *Off the Beaten Track*, edited and translated by Julian Young and Kenneth Haynes, 1-56. Cambridge: Cambridge University Press.

———. *Über den Humanismus*. 1947. Frankfurt am Main: Vittorio Klostermann, 1949.

"Brief über den Humanismus." In GA. Vol. 9, *Wegmarken*, 313–64.

"Letter on 'Humanism.'" In *Pathmarks*. Edited by William McNeill, 239–76. Cambridge: Cambridge University Press, 1998.

———. "Überwindung der Metaphysik." In *Vorträge und Aufsätze*, 67–97. Pfullingen: G. Neske, 1954.

"Überwindung der Metaphysik." In GA. Vol. 7, 67–98.

"Overcoming Metaphysics." In *The End of Philosophy*. Translated by Joan Stambaugh, 84–110. London: Souvenir Press, 1975.

———. *Vom Wesen der Wahrheit* (1943). 2nd ed. Frankfurt am Main: Vittorio Klostermann, 1949.

"Vom Wesen der Wahrheit." In GA. Vol. 9, 177–202.

"On the Essence of Truth." In *Pathmarks*, 136–54.

———. *Vom Wesen des Grundes* (1929). 4th ed. Frankfurt am Main: Vittorio Klostermann, 1955.

"Vom Wesen des Grundes." In GA. Vol. 9, 123–75.

"On the Essence of Ground." In *Pathmarks*, 97–135.

———. *Was heißt Denken?* Tübingen: Niemeyer, 1954.

GA. Vol. 8, *Was heißt Denken?*

What is Called Thinking? Translated by J. Glenn Gray and Fred D. Wieck. New York: Harper & Row, 1968.

———. *Was ist das – die Philosophie?* Pfullingen: G. Neske, 1956.

"Was ist das – die Philosophie?" In GA. Vol. 11, 3–26.

What is Philosophy? Translated and introduced by Jean T. Wilde and William Kluback. German-English edition. Lanham, MD: Rowman & Littlefield, 1956.

———. *Was ist Metaphysik?* 5th ed. Frankfurt am Main: Vittorio Klostermann, 1949.

"Was ist Metaphysik?" In GA. Vol. 9, 103–22.

"Einleitung zu 'Was ist Metaphysik?'" In GA. Vol. 9, 365–83.

"Nachwort zu 'Was ist Metaphysik?'" In GA. Vol. 9, 303–12.

"What is Metaphysics?" In *Pathmarks*, 82–96.

"Introduction to 'What is Metaphysics?'" In *Pathmarks*, 277–90.

"Postscript to 'What is Metaphysics?'" In *Pathmarks*, 231–38.

———. "Die Zeit des Weltbildes." In *Holzwege*, 69–104. Frankfurt am Main: Vittorio Klostermann, 1950.

"Die Zeit des Weltbildes." In GA. Vol. 5, 75–113.

"The Age of the World Picture." In *The Question Concerning Technology and Other Essays*, translated by William Lovitt, 115–54. New York: Harper & Row, 1977.

———. *Zur Seinsfrage*. Frankfurt am Main: Vittorio Klostermann, 1956.

"Zur Seinsfrage." In GA. Vol 9, 385–426.

"On the Question of Being." In *Pathmarks*, 291–322.

Heraclitus. "Heraclitus." In *Early Greek Philosophy*, translated and edited by Jonathan Barnes. London: Penguin, 1987.

Hölderlin, Friedrich. "Fragment von Hyperion." In *Sämtliche Werke und Briefe in drei Bänden*. Edited by Jochen Schmidt. Vol. 2, *Hyperion. Empedokles. Aufsätze. Übersetzungen.*

172 ———. "Hyperion oder Der Eremit in Griechenland." in *Sämtliche Werke und Briefe in drei Bänden*. Vol. 2, *Hyperion. Empedokles. Aufsätze. Übersetzungen*. Frankfurt am Main: Deutscher Klassiker Verlag, 1994.

    "Hyperion, or the Hermit in Greece." In *Hyperion and Selected Poems*, edited by Eric L. Santner, 1–133. New York: Continuum, 1990.

Jaspers, Karl. *Vernunft und Widervernunft in unserer Zeit*. Munich: Piper, 1950.

    Reason and Anti-Reason in Our Times. Translated by S. Goodman. London: SCM Press, 1952.

Kant, Immanuel. "Metaphysische Anfangsgründe der Naturwissenschaft." In *Kants Gesammelte Schriften*. Vol 4, 465–565. Berlin: Georg Reimer, 1903.

    "Metaphysical Foundations of Natural Science." In *Theoretical Philosophy After 1781*. Edited by Henry Allison and Peter Heath, 171–270. Cambridge: Cambridge University Press, 2002.

Lenin, Vladimir. "The Three Sources and Three Component Parts of Marxism." In *Collected Works*. Vol. 19. Moscow: Progress Publishers, 1977.

Lukács, Georg. *Existentialismus oder Marxismus?* Berlin: Aufbau, 1951.

———. *Geschichte und Klassenbewußtsein*: *Studien über marxistische Dialektik*. Berlin: Malik, 1923.

    History and Class Consciousness. Translated by Rodney Livingstone. London: Merlin, 1971.

———. "Heidegger Redivivus." *Sinn und Form*, no. 3 (1949): 37–62.

———. *Die Seele und die Formen*. Berlin: Egon Fleischel & Co., *1911*.

    Soul and Forms. Edited by John T. Saunders and Katie Terezakis. Translated by Anna Bostock. New York: Columbia University Press, 2010.

———. *Die Zerstörung der Vernunft*. Berlin: Aufbau, 1954.

    The Destruction of Reason. Translated by Peter R. Palmer. London: Merlin, 1980.

Marx, Karl. "Aus der Doktordissertation" (1840). In *Die Frühschriften*. Edited by Siegfried Landshut, 12–19. Stuttgart: Kröner, 1953.

    "Notes to the doctoral Dissertation" (1839–41). In *Writings of the Young Marx on Philosophy and Society*. Edited by Loyd D. Easton and Kurt H. Guddat, 52–53. New York: Doubleday, 1967.

———. *Die deutsche Ideologie* (1845/46). Berlin: Dietz, 1953.

    The German Ideology. Moscow: Progress Publishers, 1964.

———. *Das Kapital: Kritik der politischen Ökonomie; Erster Band*. 1867. Berlin: Dietz 1955.

    Capital: A Critique of Political Economy; Volume I. Translated by Ben Fowkes. London: Penguin in association with New Left Review, 1976.

———. *Kritik des Gothaer Programms* (1875). Berlin: Neuer Weg, 1946.

    "Critique of the Gotha Programme." In *The First International and After: Political Writings Volume 3*, edited by David Fernbach. London: Penguin in association with New Left Review, 1974.

———. *Misère de la philosophie*. 1847. Paris: A. Costes 1950.

    The Poverty of Philosophy. London: Martin Laurence Limited, 1937.

———. "Nationalökonomie und Philosophie." In *Die Frühschriften*, 225–316.

"Economic and Philosophical Manuscripts." In *Early Wiritings*, 279–400.

———. "Thesen über Feuerbach." In *Die Frühschriften*, 339–41.

"Concerning Feuerbach." *Early Wiritings*, 421–23.

———. Zur *Kritik der politischen Ökonomie*. Berlin: Dietz, 1951.

Grundrisse: Foundations of the Critique of Political Economy (Rough Draft), translated by Martin Nicolaus. London: Penguin in association with New Left Review, 1973.

Nietzsche, Friedrich. "Also sprach Zarathustra II." In KSA. Vol. 4, 103–90.

"Thus Spoke Zarathustra: Second Part." Edited and translated by Walter Kaufmann. In *The Portable Nietzsche*. New York: Penguin, 1954.

———. "Die fröhliche Wissenschaft." In KSA. Vol. 3, 334–651

The Gay Science. Translated by Walter Kaufmann. New York: Vintage, 1974.

———. *Großoktavausgabe*. Edited by Elisabeth Förster-Nietzsche et. al. Leipzig: Kröner, 1894–1904.

KSA. Vol. 9, *Nachgelassene Fragmente 1880–1882*.

KSA. Vol. 11, *Nachgelassene Fragmente 1884–1885*.

———. "Jenseits von Gut und Böse." In KSA. Vol. 5, 9–243.

"Beyond Good and Evil." In *Beyond Good and Evil/On the Genealogy of Morality*, translated by Adrian Del Caro. Stanford: Stanford University Press, 2014.

Novalis. *Werke, Tagebücher und Briefe Friedrich von Hardenbergs*. Edited by Hans-Joachim Mähl and Richard Samuel. 2 vols. Munich: Hanser 1978.

Plato. *Cratylus*. Translated by Harold N. Fowler. Greek-English edition. Cambridge, MA: Harvard University Press, 1921.

Proudhon, Pierre-Joseph. *Système des contradictions* économiques *ou Philosophie de la misère*. Paris: Guillaumin & Cie, 1846.

System of Economic Contradictions: Or, The Philosophy of Misery. Translated by Benjamin R. Tucker. Boston: Benj. R. Tucker, 1888.

Remarks

Marx and Heidegger: Guides to a Future Way of Thought

The largest part of this text was presented during a two-hour guest lecture on 12 July 1957 in the Auditorium Maximum of the Freie Universität Berlin; on 13 July 1957, a colloquium was held on the same subject matter in the Philosophical Seminar; and on 27 January 1966, the latter was repeated in the Neue Aula of the Universität Tübingen.

Theses on Marx: Concerning the Critique of Philosophy, of Political Economics, and of Politics

First published in *Die Neue Rundschau*, Frankfurt am Main, Issue 2, 1961. A French version appeared in the journal *Arguments*, Paris, No. 7, 1958 and was included in the book *Vers la pensée planétaire. Le devenir-pensée du monde et de devenir-monde de la pensée*, Collection "Arguments" (Paris, Éditions de Minuit, 1964).[1]

Concerning the Experience of the World: On Heidegger

Unpublished.

The Planetary: A World History of Technology

A lecture held on 13 June 1956 at the Kant Society in Berlin. It was repeated on 21 June 1957 at the Universität Freiburg im Breisgau in the context of general studies as well as on 22 June 1957 at the Fürstabt-Gerbert-Haus for students of the Saint Blasien University Sanatorium (in the Black Forest). Certain parts were not presented at the respective events.

---

1   EN: The French text was previously translated by N. Georgcpoulos as "Theses on Marx," in *Continuity and Change in Marxism*, ed. N. Fischer, N. Georgopoulos and L. Patsouras (Atlantic Highlands, NJ: Humanities Press, 1982), 66–70.

Twelve Fragmentary Propositions Concerning the Issue of Revolutionary Praxis

First published in the Tübingen Student Newspaper *Notizen*, November 1965.

A Discussion about Science

Closing discussion conducted with the classical philologist Jean Bollack following the radio series "Problems and Performances of Modern French Science" in the scientific evening program of Sender Freies Berlin on 6 August 1957. The nuclear researcher Delcroix, the physiologist Cahn, the geographer Roncayolo, the ethnologist Lévi-Strauss, the economic analyst Sauvy, and the philosopher-anthropologists Ricoeur and Foucault had spoken about their work, their successes, and their doubts.[2]

2   EN: While Claude Lévi-Strauss, Paul Ricoeur and Michel Foucault are well-known, the other figures are more obscure. These are likely to be Jean-Loup Delcroix, Théophile Cahn, Marcel Roncayolo, and Alfred Sauvy. As far as I am aware the other conversations have not been published.

Bibliography of Works by Kostas Axelos and English Translations

Φιλοσοφικές Δοκιμές [Philosophical Essays]. Athens: Papazeses, 1952.

Marx penseur de la technique. Paris: Les Éditions de Minuit, 1961.

> *Alienation, Praxis and Techne in the Thought of Karl Marx*. Translated by Ronald Bruzina. Austin: University of Texas Press, 1976.

Héraclite et la philosophie. Paris: Les Éditions de Minuit, 1962.

Vers la pensée planétaire. Paris: Les Éditions de Minuit, 1964.

> Part of "Introduction à la pensée planétaire" (20–24) and the whole of "L'interlude" (321–28) translated by Sally Hess. "Planetary Interlude." *Yale French Studies* 41 (1968): 6–18.
>
> "Thèse sur Marx" (172–77) translated by N. Georgepoulos. "Theses on Marx." In *Continuity and Change in Marxism*. Edited by N. Fischer, N. Georgopoulos and L. Patsouras, 66–70. Atlantic Highlands, NJ: Humanities Press, 1982. See also the first chapter in part two of the current volume.

Einführung in ein künftiges Denken: Über Marx und Heidegger. Tübingen: Niemeyer, 1966.

> *Introduction to a Future Way of Thought: On Marx and Heidegger*. Edited by Stuart Elden. Translated by Kenneth Mills. Lüneburg: Meson Press, 2015.

Le Jeu du monde. Paris: Les Éditions de Minuit, 1969.

Arguments d'une recherche. Paris: Les Éditions de Minuit, 1969.

Pour une éthique problématique, Paris: Les Éditions de Minuit, 1972.

Entretiens, réels, imaginaires et avec soi-même. Montpellier: Fata Morgana, 1973.

Horizons du monde. Paris: Les Éditions de Minuit, 1974.

> "Le jeu de l'ensembles des ensembles" (77–84) translated by Beverly Livingston. "The Set's Game-Play of Sets." *Yale French Studies* 58 (1979): 95–101; and by R. E. Chumbley. "Play as the System of Systems." *Sub-Stance* 8, no. 4 (1979): 20–24.
>
> "Marx, Freud et les taches de la pensée future" (87–99), translated by Sally Bradshaw, "Marx, Freud, and the Undertakings of Thought in the Future," *Diogenes* 72 (1970): 96–111.

Contribution à la logique. Paris: Les Éditions de Minuit, 1977.

Problèmes de l'enjeu. Paris: Les Éditions de Minuit, 1979.

Systématique ouverte. Paris: Les Éditions de Minuit, 1984.

> "Le monde: L'être en devenir de la totalité. Cela" (40–54) translated by Gerald Moore. "The World: Being Becoming Totality." *Environment and Planning D: Society and Space* 24, no. 5 (2006): 643–51.

Métamorphoses: Clotûre-ouverture. Paris: Les Éditions de Minuit, 1991.

L'Errance érotique. Brussels: La Lettre volée, 1992. (Reprint of *Vers la pensée plan-étaire*, 273–96.)

Από το εργαστήρι της σκέψης [From the Workshop of Thought]. Athens: Hestia, 1992.

178   *Γιατί σκεφτόμαστε; Τι να πράξουμε* [Why Think? What is to be Done?]. Athens:
      Nepheli, 1993.

  Lettres à un jeune penseur. Paris: Les Éditions de Minuit, 1996.

  Métamorphoses. Paris: Hachette-Littérature "Pluriel," 1996.

  Notices "autobiographiques." Paris: Les Éditions de Minuit, 1997.

  Ce questionnement. Paris: Les Éditions de Minuit, 2001.

  Η εποχη το υπατο διακυβευμα [The Challenge of the Epoch]. Athens: Nepheli, 2002.

  Réponses énigmatiques. Paris: Les Éditions de Minuit, 2005.

  Ce qui advient. Fragments d'une approche. Paris: Encre marine, 2009.

  Το άνοιγμα στο επερχόμενο και Το αίνιγμα της τέχνης [The Opening and the Future:
      The Enigma of Art]. Athens: Nepheli, 2009.

  En quête de l'impensé, postface de Servanne Jollivet. Paris: Encre marine, 2012.

  Le Destin de la Grèce modern. Paris: Encre marine, 2013.

English Interviews

"Mondialisation without the World." Interview with Stuart Elden. *Radical Philosophy*,
    no. 130 (March/April 2005): 25–28.

"For Marx and Marxism: An Interview with Kostas Axelos." By Christos Memo. *Thesis
    Eleven*, no. 98 (August 2009): 129–139.

www.ingramcontent.com/pod-product-compliance
Lightning Source LLC
Chambersburg PA
CBHW031153020426
42333CB00013B/639